Understanding Leicestershire and Rutland Place-names

Jill Bourne

Heart of Albion Press

UNDERSTANDING LEICESTERSHIRE AND RUTLAND PLACE-NAMES

Jill Bourne

Cover photomontage by Bob Trubshaw

ISBN 1 872883 71 0

Heart of Albion Press
2 Cross Hill Close, Wymeswold
Loughborough, LE12 6UJ

albion@indigogroup.co.uk

Visit our Web site: www.hoap.co.uk

Printed in the UK by Booksprint

Contents:

List of maps:

Maps prepared by Dr Anne Tarver.

THE LANGUAGES AND CHRONOLOGY OF ENGLISH PLACE-NAMES

Place-names have an amazing capacity for survival. The oldest written records of Leicestershire and Rutland place-names date back to the eighth century, although most were only written down for the first time in the Domesday Book of 1086. Undoubtedly every name is older, perhaps much older, than the date of the first surviving record. The various layers of place-names reflect the migrations, conquests, and settlements of the people who came here and the different languages they spoke.

Non-Indo-European language(s)

In England six successive layers of languages can be distinguished in place-names. The earliest language(s) are those of the people who were here before the coming of the Celts. Almost nothing is known about these languages but their existence can be assumed by the survival of a few words, mainly river names, which have no basis in any Indo-European language. The River Soar is probably one of these pre-Celtic survivals.

Celtic

During the last four centuries BC Iron Age people migrated and settled in this country. Their language and dialects are part of the Indo-European family of languages. Celtic is divided into two branches, Goidelic and Gaelic. Gaelic is the basis of Irish and Scottish Gaelic, and Goidelic is the basis of Manx and Brittonic, or British, to which Welsh and Cornish belong. British was the language which was spoken throughout mainland Britain, up to the Forth and Clyde, until the arrival of the Anglo-Saxons who borrowed some of these Celtic words when they came to Britain from the fifth century onwards. These Celtic names can still be found all over England, a few survive in the east, with more and more surviving crossing the country westward until, in Cornwall and Wales, they predominate.

Latin

Latin is also an Indo-European language which was in use in Britain from the mid-first to mid-fifth centuries during the time of the Roman occupation. Latin left very little mark on place-names as it was spoken and written by the Roman administrators and the army but not usually by the native British, particularly in the countryside where the Celtic language would have persisted. Over the four centuries of

Roman occupation presumably many of the British would have picked up enough Latin to get by and no doubt some might have adopted the language in order to better themselves by working with the ruling elite. The British elite would have been highly romanised and may well have sought an education based on Latin for their children. Latin place-names are rare in Britain but there are many British and Anglo-Saxon names which contain Latin words.

Old English
This is the term used for the language which was spoken by the Anglo-Saxons who began to come to Britain to invade and colonise after the decline and fall of Roman rule here in the first half of the fifth century AD. Anglo-Saxon settlement would have spread slowly from the east of the country until by the ninth century all of what is now England was occupied, with the exception of some areas along the Welsh border and Cornwall. The new settlers were Angles, Saxons, and Jutes who had migrated from northern Europe. They spoke a common language, now usually called Old English (abbreviated to OE). It was a Germanic language which belonged to a different branch of Indo-European languages from Celtic. The change of population and language in Britain led to the introduction of thousands of Anglo-Saxon place-names, the majority of which are the names in use today although, in most cases, in a much changed form. This virtual total change of language must represent the political domination of a large number of Anglo-Saxons over the indigenous Celtic-speaking British.

Old Scandinavian
The Viking invasions, and subsequent settlement of Britain, took place during the ninth, tenth, and eleventh centuries. The languages they spoke, Old Danish and Old Norse, were very similar, and were, like Old English, part of the Germanic branch of the Indo-European group of languages. In this book the term Old Scandinavian (abbreviated to OScand.) is used for all Scandinavian words whether Danish, Norwegian, Swedish, or Icelandic. Scandinavian place-names are largely found north of the Thames in central and eastern England and parts of the north-west, the regions of the country which took the impact of the Danish and Norwegian Viking invasions and settlement.

The armies eventually settled as is recorded in the Anglo-Saxon Chronicle and were, almost certainly followed by non military settlers. Old English and Old Norse are similar languages in many ways and the incomers would have taken over many English settlements without altering their names. Other names would have been altered when there were problems with pronunciation and some would have been deliberately renamed when Viking settlers ousted the incumbent Anglo-Saxon land-holders. The Vikings also founded and named completely new settlements.

Norman French

This was the language of William the Conqueror and his Norman followers. It is estimated that they numbered about 7,000 in the years immediately after 1066. The Normans were themselves of Germanic origin; they were Vikings, 'northmen', who ravaged northern France in the later years of the ninth century. William's Dukedom was the land given to his forebears by the French King Charles the Simple in 911 as a pact to stop the raiding. These 'northmen' did not impose their language on the people of the Dukedom; instead they adopted French. Neither did they impose their language on the defeated English after 1066, although Norman-French did have a considerable influence on the development of the English Language. This would bring about language change in the same way as does the new language of a large number of peasant settlers. This explains the relatively few surviving Latin place names. The pre-eminence of Old English as the language of this country must reflect the high numbers of Anglo-Saxon invaders and settlers in the centuries after the end of Roman rule.

THE TECHNIQUE OF PLACE-NAME STUDY

In order to deduce the origins and original meaning of a place-name, we have to make, from the earliest records down to the present day, a collection of the forms or spellings of the place-name, tracing the history of the spelling of the name from its current form back to its earliest recorded one which, in most cases, is the name recorded in the Domesday Book of 1086. This collection of the various forms of a name shows how the name has developed during its recorded history. The majority of these names are over 900 years old, and many will have had a longer or shorter history before the name was first recorded. Some go back to the Viking invasion and settlement in the late ninth century. Others, indeed most in Leicestershire and Rutland, date from the Anglo-Saxon period; some go back to Romano-British times, and a few are even earlier.

Very few names are recorded in pre-Domesday Book documents but it is only from these recorded names with their Old English spellings that an etymology can be deduced. This collection of spellings enables us to trace the linguistic development of the name and along with our knowledge of the ways in which the English language developed in the Anglo-Saxon period, this leads to the form the name would have had when it was first used. Although this form is of course hypothetical it must explain the development of our collection of names and it must be the one that would have been expected from all the collected evidence.

Some of the important sources for early spellings of place-names

c. AD 150 Geography of Ptolemy
c. AD 300 Antonine Itinerary
c. AD 670 Cosmography of Ravenna

These three sources mainly record the names of Roman towns and fortifications; they also record river names, albeit in a Latinised form, which are mainly of Celtic origin.

The Anglo-Saxon Chronicle
This document appears to have been first drawn up towards the end of the ninth century although it includes material from much earlier times and sources. Most of the entries are short; there are only one or two entries for each year and none for some.

Land charters
Nationally almost 2,000 charters survive from the Anglo-Saxon period most of which contain references to places. In the areas of the country where the Vikings settled, such as Leicestershire and Rutland, almost no charters have survived. This must reflect the deliberate destruction of records by the incomers who were both pagan and illiterate. Some of these early charters are original but most are copies from the twelfth to fourteenth centuries. Some may contain errors of transcription along with some deliberate changes particularly with names which the copyist may have 'modernised'.

1086 Domesday Book
For the vast majority of places the entry in Domesday Book is the first time they appear in the written record. Domesday Book was compiled under the orders of William the Conqueror and intended as a record of lands and their value. Norman scribes travelled the country gathering the information. Their language was Norman French so they would not have any knowledge of the language of the

people they were questioning. They would have tried to accurately record the sound of the names but would have used the nearest equivalent spelling in Norman French. Domesday spellings must be used with caution although in most cases it is possible to produce a satisfactory explanation.

Post-Domesday documents up to 1500

These are mainly official documents such as Pipe Rolls, Charter Rolls, Close Rolls, to name but a few of the hundreds of sources that could contain early forms of names. There are also thousands of charters which might include useful evidence, some of these are to be found, along with the official documents, at the Public Record Office. Others might be lodged in County Record Offices and some are in the private collections of large land-owners such as the Dukes of Rutland. These source materials are too numerous to list here but from this type of document most place-name forms are gathered and when they are all listed provide a body of names from which specialists in the early languages of the settlers can deduce the probable meaning of a name.

Place-name research has come a long way in the past thirty or so years and we are now in the fortunate position that all the early forms of almost all names of settlements in England have been researched and the etymology and philology of the names determined. This means that it is now possible for local historians, archaeologists, geographers or anyone interested to use place-names with confidence. We are grateful for the work of scholars like Dr Margaret Gelling, Professor Kenneth Cameron and Dr Barrie Cox, to name but three of several leading scholars, for making this possible; see further reading on page 124.

PLACE-NAME ELEMENTS

The elements that make up place-names can, broadly speaking, be categorised into two main groups, habitative and topographical. Habitative elements indicate a settlement of some sort. Topographical elements are words which describe aspects of the landscape, both natural and created.

SOME OF THE MORE COMMON HABITATIVE ELEMENTS

Anglo-Saxon words

OE *ham*

The modern word 'home' derives from *ham*, which comes from the German base *Xaim* and also from the Old Scandinavian word *heim* meaning 'home'. *Ham* commonly occurs as a second or final element in a place-name and usually means 'village' but it can also mean 'house, manor, estate', or even sometimes 'a religious household'. *Ham* occurs fairly early in the period of Anglo-Saxon settlement of England and is often combined with a personal name as the first element.

Care must be taken with this element as *ham* is easily confused with *hamm* which means 'land in a river bend, river meadow, dry ground in a marsh'. All early spellings need to be examined to see which it is and if there is any doubt the topography must be examined on the ground.

OE *ing ingas ingaham ingatun*

All of these elements have a variety of uses. *-inga* (singular) and *-ingas* (plural) when combined with a personal name indicates an association of people depending on a common leader, 'the people of', and living in a particular place. It is added to both personal names and topographical terms. Such a group of 'folk' might have been partially or largely kin although blood relationship was not necessarily the only basis of their connection. An invading or colonising group might have originated from a common settlement in the homelands comprising both neighbours and extended family.

-ington represents the possessive form ('belonging to') of *-inga*, *-ingas* as in *-ingaham*, or is used as a connective word linking the first element of a place-name to a second, meaning 'pertaining to someone or something'.

OE *tun*

This is the commonest element in English place-names and is the origin of the English word 'town'. There are only six examples of *tun* found in the written records before AD 750 and when it does occur it is used in a highly specific way. By the second half of the eighth century its use steadily increases until it becomes the most common element in English place-names. In compound names *-tun* always occurs as the final element.

The word *tun* probably derives from the Celtic-Germanic word *tunaz* or *turnam* meaning 'a fence or hedge'. It then develops in Old English into 'an enclosed piece of land, a building, a farmstead with its own enclosed piece of land, a hamlet, a village, a manor, an estate'.

OE *leah*

This element derives from an Indo-European root *louq* which is related to the root of the modern noun 'light' and the Latin *lucus* meaning 'a grove'. It survives as the modern 'lea' and was used to refer to both 'woodland' and to a 'clearing in woodland' the latter being the most likely meaning when used in place-names. The clearing may have been a cultivated woodland clearing used for pastoral or arable farming. *Leah* later developed into 'open land' or 'meadow land'. The element *-leah* is, as would be expected, most commonly found in woodland areas.

OE *worth*

The element *worth* broadly speaking means 'an enclosure' of some sort. What was enclosed varied from a single building or farmstead to a whole settlement such as Tamworth, the most important royal settlement of the kings of Mercia.

Scandinavian elements

Many place-names are in the language spoken by the Vikings, known as Old Norse or Old Scandinavian. The Vikings who came to Britain were from Norway and Denmark. The Swedish Vikings, who had no outlet to the sea in the west, went eastwards. Although the basic language was the same amongst the various Viking groups, there were considerable differences in dialect which led to differences in the terminology of place-names between the areas settled by Norwegians and those by Danes. The vocabulary of Old Scandinavian is similar to Old English. This can sometimes lead to difficulties in determining the origin of a word and is often further complicated by the Scandinavianisation of Old English words.

Hybrid names

Commonly recurring name types that reflect Viking influence are 'hybrid names'. These are where a Scandinavian specific, most frequently a personal name, is compounded with the Old English (OE) generic **tun**. These names are sometimes known as 'Grimston hybrids' where a Viking male personal name **Grimr** (or any other Viking personal name) is combined with the OE **tun**. These names are interesting in that they show that already existing Anglo-Saxon settlements were appropriated by Viking incomers who kept the generic **tun** and added their own names. The adding of a personal name but keeping **tun** could well have been a deliberate act to emphasise the new overlordship. Many of these men whose names are enshrined in the place-names we use today would have fought in the 'great heathen army' which is referred to numerous times in the Anglo-Saxon Chronicle during the ninth century. Some hybrid names may date from the eleventh century and mark the transference of land to the followers of Knut the Great (King Canute) or his sons in the eleventh century.

-by

This is the commonest Scandinavian habitative element in English place-names. It is found in all parts of the Danelaw, especially in Lincolnshire, Leicestershire, and the North Riding of Yorkshire but not in County Durham, Northumberland and Rutland. It is also common in the north-west

In England **-by** seems to have been used for any kind of settlement from a farmstead to a town, but most commonly a farmstead or village. The word has the same root as the modern words 'bower' or 'build' which derive from the Germanic **bu** 'to dwell' or 'cultivate'. This element had a long history back in the Scandinavian homelands where its meaning was either 'a new cultivation', 'a secondary settlement from an older existing settlement', or an 'outlying isolated farm'. This is not the way the word was used in the Viking settlement names in this country where **-by** had the meaning of 'farmstead or village' but with no suggestion of dependency or of newly cultivated land. In this country over ninety percent of place-names with the suffix **-by** have a personal name as their first element; this is in sharp contrast to the homelands where less than ten percent of place-names contain a personal name. The element **-by** is associated with a period of Scandinavian settlement which is slightly later than the hybrids names and probably reflects settlement by the rank and file soldiery along with non-military migrants.

thorp

This is the second most frequently occurring Scandinavian habitative element in English place-names. It is Danish in origin and rarely found where the Viking settlers were mainly Norwegian. The word denotes an outlying, secondary, dependent settlement which could have been either a single farmstead or a hamlet or villages. It represents the final stage of Scandinavian settlement when there had been considerable integration between the Anglo-Saxons and the Scandinavians. In fact the personal names compounded with *thorp* are both Sandinavian and Anglo-Saxon.

Thorps are not an homogeneous group. Some have developed over time into large villages but most have remained relatively small and undistinguished. Nearly half are shrunken or lost settlements and only half are listed separately in Domesday Book, confirming their original status as secondary settlements founded on less fertile soils.

There is an Old English word *throp* which also means 'an outlying farm or village'. This can be confused with the Old Scandinavian *thorp*; fortunately there are only a very few *throps* compared with *thorps* and none of these occur with any certainty in either Leicestershire or Rutland. Only an extensive analysis of all the early spellings can distinguish between these two elements.

TOPOGRAPHICAL PLACE-NAMES

Topographical names define settlements by describing their physical surroundings. This is in contrast with the habitative place names already discussed, although the two categories overlap. Most place-names are compounds of two, sometimes three elements (Blackfordby) and often a name containing a habitative word is qualified by a landscape term. Bottesford, Ashby, Hambledon, Burton, Coleorton, Belton are just a few of the scores of these compound names in Leicestershire and Rutland. In spite of a considerable overlap of the two categories of name the distinction is clear. In topographical names the main emphasis is the geographical feature not on the type of settlement. Many place-names contain only topographical terms such as Tixover, Barrowden, Cranoe and Pickwell.

Perhaps the most significant aspect of topographical names, particularly for major landform terms (hills, valleys) is that they form a country-wide system of naming, a remarkable fact given the difficulties and slowness of movement around the country. It is unlikely that there was an organised force behind this consistency; the probable explanation may lie in the needs of travellers to get to a destination and to pass that information on to others. The naming of places would have evolved over several generations until a descriptive name became attached to a feature.

For a detailed, up to date and accessible discussion of topographical place-names see *The Landscape of Place-names* by Margaret Gelling and Ann Cole (Shaun Tyas 2000).

THE VIKINGS IN LEICESTERSHIRE AND RUTLAND

In 877 AD the Anglo-Saxon Chronicle records that *'tha on herfeste geffor se here on Myrcena land and hit gedaeldon sum'*, 'then in the harvest season the army went away into Mercia and shared out some of it'. This documentary reference is clear and yet the evidence for the presence of Vikings in Leicestershire and Rutland is slight. Few artefacts have been found, there are just a handful of documentary references and few excavations have produced Viking evidence. This leaves place-names as the only significant body of evidence available to us of Viking settlement in the two counties. The names occur in such abundance that there can be no doubt of there being a significant Viking presence here.

Viking place-names are scattered, in small clusters, over most parts of Leicestershire. They stop abruptly, with the exception of a few stragglers which can be explained by specific local conditions, at Watling Street which is the Leicestershire county boundary with Warwickshire. There is also a marked falling off of Scandinavian names to the west of the Fosse Way north of Leicester.

In sharp contrast Rutland, although surrounded by areas of dense Viking settlement, has almost no primary Scandinavian names. There are just two hybrid names, Glaston and Normanton and they are in a unique category in the county. Glaston is the Norwegian personal name **Glathr + tun,** and Normanton is **northmanna + ton,** 'the Norwegian's settlement'. These two places appear to have been settled at a time when Rutland was under the control of the Norwegian Vikings in York and may represent a political decision to have representation in the heart of the otherwise Anglo-Saxon county.

There are no names in **-by** in Rutland. There are 14 **thorps,** of which only five are recorded in Domesday Book, and are all late-developed secondary dependent settlements sited on or near the county boundaries. This relative lack of a Scandinavian presence in Rutland suggests that a political explanation lies behind this phenomenon, probably in the form of a treaty between Ceolwulf, King of the Mercians, and the Viking leaders who were involved in the division of the Mercian kingdom in 877. After the reconquest of the Danelaw by Edward the Elder between 918 and 921 Rutland seems to have passed to a unified England, becoming part of (perhaps reverting back to) the dower-land of the Anglo-Saxon queens; the name of one, Queen Edith, is enshrined in the name Edith Weston.

The Vikings in the Wreake valley

The most striking distribution of Viking names in Leicestershire is in the north-east of the County particularly along the Wreake Valley and its tributaries. Here almost three-quarters of the place-names are either wholly or partly Viking, or show evidence of Scandinavian influence (as in Melton and Scalford). Names which comprise a Viking personal name plus the Old English *tun* may have been coined slightly earlier than the names which contain the suffix *by*. The *bys* might represent a slightly later breaking up of earlier large Anglo-Saxon estates into smaller units. The density of Viking settlement in the East Midlands, Lincolnshire and East Yorkshire represents a larger group of people than would have served in the Viking armies over the years. There must have been considerable colonisation from Scandinavia in the years subsequent to the armies breaking up. These colonisers probably came overland into the region. The persistence of significant numbers of Anglo-Saxon names suggests that although there would have been some displacement of the Anglo-Saxon population it was by no means total. Geological evidence also indicates that many of the Scandinavian-named settlements were on the less desirable soils.

The name of the River Wreake is of particular interest and probably unique. It is called the Eye in its upper reaches until it reaches Melton where its name changes to the Wreake, meaning 'to bend and twist', until its confluence with the Soar. This rare name-change must reflect the overwhelmingly Viking character of the population. From Melton to the River Soar both banks of the river, and its tributaries, appear to have been settled by Vikings whereas there are no Viking names to south of the river from Melton eastwards to the Lincolnshire boundary. The almost total absence of primary Viking names in Rutland and the three parishes south of the river Wreake which border on to Rutland suggests that this separation is not one of chance. Perhaps this marked division could be connected with the suggested treaty that applied in Rutland. Edmondthorpe has to be discounted as it is a secondary dependent

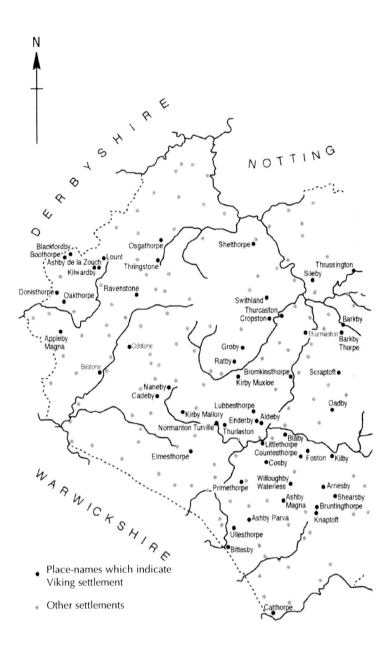

Place-names which indicate Viking settlement

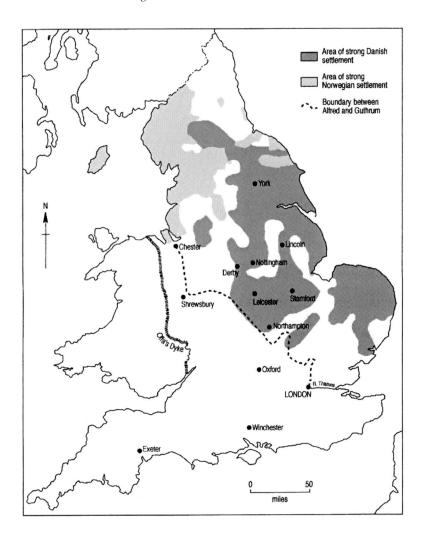

England showing the broad areas of Danish Viking and Norwegian Viking settlement

place settled from Wymondham and, given the fact that it is listed in Domesday Book, it could well be an OE **throp** rather than an OScand. **thorp**.

Framland Wapentake, in which most of this dense Viking settlement lies, is also a Scandinavian word. The moot site lay two miles north of Melton Mowbray, at present day Great Framlands, on the ancient routeway which runs south from the Vale of Belvoir to cross the Wreake at Melton Mowbray.

DESERTED MEDIEVAL VILLAGES OF LEICESTERSHIRE AND RUTLAND

In England there are over 2,000 settlements that were deserted, mainly during the Medieval period. Most of these are in the Midlands, Eastern England, the lowlands of the South and South-West and the plains of the North, all areas where there were extensive ploughlands of 'open-field' farming. The terms 'deserted', or 'lost' or 'shrunken' apply to any settlement for which there is surviving evidence that it existed in medieval times but where now nothing, or virtually nothing, remains, perhaps at most a farm, a house, or a church. Great Stretton is such a place. In Leicestershire and Rutland there are 80 deserted or shrunken villages or hamlets, 70 of which are still completely deserted.

Evidence on the ground
The best time to look at the site of a lost village is in the early evening in late autumn, winter, and early spring when the grass is short and the shadows long. Another perfect time to pick out the shape of a lost village is after a light sprinkling of snow or a heavy frost.

Although deserted settlements are varied in their plan, there are features which can be seen on most sites. A sunken, hollow, 'main' street can often be seen running through the centre of the settlement. This hollowing would have been caused by erosion brought about over the centuries by the passage of people, animals, and carts. The road would, at most, have had a few cobbles and stones pushed into it to hold the surface together. Side roads can sometimes be seen with a back lane running behind the gardens of the cottage dwellings that lay along the streets. On some sites a bank and ditch surrounds the settlement; this may have had a defensive function but was more likely to have been intended to keep domestic animals away from the

Sites of deserted medieval settlements.

Also included are places which were deserted but
have grown up again on the same site or nearby.

crops and to keep wild animals, such as wolves, out of the village at night.

There is often a moat on the site of a manor house; occasionally the remains of the manor house itself can be seen incorporated into the structure of later buildings. The domestic building most likely to have survived desertion is the manor house because of its superior construction compared with the cottage dwellings. All that usually remains of these humble cottages, which were usually built of wattle and daub, are grassy mounds which cover the rubble of their foundations. Occasionally the church has survived, as at Great Stretton, Tixover, and Wistow. Sometimes just a fragment of the structure of the church survives, as at Pickworth and Knaptoft, but usually the building stone has been robbed to build new buildings elsewhere. A typical site may also have a mound on which a windmill stood, there might have also been a dovecote and fishponds but, without either maps or a comprehensive excavation of a site, it can be difficult to sort out the grassy mounds one from the other.

Causes of de-population and desertion

The Black Death of 1348–9 is usually regarded as the great divide between the expansion of the English landscape and population and the subsequent century and a half of economic retrenchment. There were four visitations of the bubonic plague during the second half of the fourteenth century by which time between a third and a half of the population had died as a direct result of the plague. Probably no village or hamlet avoided the plague completely, although some escaped lightly while others appear to have been wiped out in their entirety, perhaps because the survivors of the plague moved to other places.

For over half a century before the Black Death, the country appears to have been stagnating, even declining, economically. There had been a steady expansion of population during the twelfth and thirteenth centuries and the population might have been seven million at the beginning of the fourteenth century. A population this large would have been difficult to support, given the farming practices of the time, without an economic recession beginning to bite. To further exacerbate the situation the early fourteenth century experienced a succession of cold rainy summers followed by poor harvests and extremely cold winters, all of which would have contributed to a weakening of the population making them vulnerable to the terrible effects of bubonic plague.

There is evidence that some settlements were already marginal, or even abandoned, before the Black Death but after 1400, with a population reduced by as much as a half, many more settlements were struggling to remain viable. A few settlements disappeared as a direct result of the plague, although of the 80 deserted settlements in Leicestershire and Rutland only eight at the most owe their demise to bubonic plague.

Traditional open-field arable farming, which was highly labour intensive, was difficult to maintain and the shortage of labour and surplus land made enclosure of fields for sheep pasture an easy process. This opportunity was widely exploited, especially in those circumstances where there was a single landowner such as at Ingarsby, where Leicester Abbey ordered the tenants to leave. Where there were multiple freeholders this process was not quite so simple and after 1600 enclosure had to be by agreement between the squire and the smaller landowners. Cold Newton and Lowesby are two such cases with the village houses there today being new developments on new sites.

Settlements were sometimes deliberately destroyed when powerful landowners decided to enlarge the park around their country houses, a fashion which began in the late fifteenth century and gathered momentum right through to the eighteenth and nineteenth centuries. When this happened, village houses, which were regarded as an eyesore, were swept away often leaving only the church standing near to the house as at Prestwold.

Hamilton and Baggrave: two local examples

Surviving records provide interesting details about the way two Leicestershire settlements became deserted villages. Medieval Hamilton, now the site of extensive housing estates and large supermarkets, stood on a 15 acre site four miles north-east of Leicester. The village was small, with about 12 families, and was a typical settlement in this part of the world at the beginning of the the fourteenth century, with its own chapel dedicated to St John the Baptist. By the time of the 1377 Poll Tax there were only four families, presumably survivors of the bubonic plague, living there. In 1423 the village passed to the Willoughbys of Wollaton who sold it on 70 years later to Thomas Keble of Humberstone. The conveyance showed that by this time the land was already completely enclosed for sheep and cattle. In 1477 Leicester Abbey records state 'we had formerly a chapel at Hamilton', suggesting that the settlement was totally deserted by the middle of the fifteenth century. Following the

Poll Tax of 1377 there had been another serious outbreak of the plague in 1389, which maybe no Hamilton villager survived.

Baggrave is a tale of enclosure which reveals the misery that countless landless peasants may have suffered. Leicester Abbey owned 216 acres of land at Baggrave. The *Domesday of Enclosures*, 1517, records that Abbot John Penny 'enclosed the messuages, cottages, and lands, with hedges and ditches. The hamlet of Baggrave is laid waste and the people have departed in tears'.

Within five generations almost 70 villages in Leicestershire and Rutland disappeared, some killed by the various visitations of the bubonic plague, others, often weakened, were swept away on a tide of economic 'progress'. For most of these settlements there is no memorial other than the grass-covered banks where the houses once stood and the hollows that that once were lanes, marking where, in that far-away time, people once lived and worked together as a community.

THE PLACE-NAMES OF LEICESTERSHIRE AND RUTLAND

ABBREVIATIONS USED

DB Domesday Book

ME Middle English (the English Language c.1100–1500)

OE Old English (the English Language c.450–1100)

OFr. Old French

OScand. Old Scandinavian (the language of the Vikings comprising Old Danish and Old Norse (Norwegian)

Nichols John Nichols *The History and Antiquities of the County of Leicester* 1795–1811

The order for each place-name is:
- The present-day form of the name, this is **emboldened**.
- Six figure O.S. map reference.
- First recorded spelling, **_emboldened, italicised_**, with date and source.
- **_Language_** and meaning of component **_elements_** of the name **_emboldened, italicised_**.
- Interpretation of the name,
- Other relevant information.

LEICESTERSHIRE

Ledcestrescire 1086 (Domesday Book)

OE *scir* 'a shire, an administrative district, a county'

The term *scir* is usually used to denote a large administrative district supervised by an alderman and later a sheriff (shire reeve). The term 'County' is a French word meaning originally 'the domain of a Count' and does not come into use in England until after the Norman Conquest.

HUNDREDS OF LEICESTERSHIRE

At the time of Domesday Leicestershire was divided into four Hundreds, Framland, Gartree, Goscote, and Guthlaxton. The Hundreds were also known by the Viking name of **Wapentac** 'Wapentake'. By the time of the Leicestershire Survey in 1130

Sites of place-names listed in the text.
See map on pages 16–17 for sites of deserted settlements.

Sparkenhoe Hundred had emerged separate from Guthlaxton and by the fourteenth century Goscote Hundred had been divided into East and West Hundred. This division is only recorded in the seventeenth century although other evidence points to the division being in the fourteenth century.

In Domesday Book all four Hundreds are referred to as **wapentac** OScand. **vapnatak** 'wapentake, division of a county'.

Framland Hundred

Franelund wapentac 1086 (Domesday Book)

Hundred 1130 (Leicestershire Survey)

OScand. **Fraena** (male personal name) + OScand. **lund** 'grove, a small wood'

Lund probably referred to a distinctive clump of trees on high ground associated with **Fraena**. In the homelands the word **lund** is often used to denote a sacred grove. The Hundred meeting place was at Great Framlands, two miles to the north of Melton Mowbray.

Gartree Hundred

Geretreu wapentack 1086 (Domesday Book)

Hundred 1130 (Leicestershire Survey)

OScand. **geirtre** 'a twisted or a spear-shaped tree'

This would probably have referred to a distinctive tree with a distorted shape. Medieval documentary evidence suggests that the meeting place lay on the (Roman) Gartree Road in the parish of Shangton, near the centre of the Hundred.

Goscote Hundred

Gosencote wapentac 1086 (Domesday Book)

Gosencote Hundred

OE **gos** 'goose' + OE **cot** 'cottage or shelter'

'A shelter for geese'. Circumstantial evidence points to Goscote Hundred being divided into East and West Hundred during the fourteenth century although the division does not appear in the record until 1604. The exact site from which the Hundred took its name is not known but may have been in the parish of Wymeswold.

Guthlaxton Hundred

Gutlacistan wapentac 1086 (Domesday Book)

Guthlac (OE male personal name) + OE **stan** 'stone'. The meeting place of Guthlaxton Hundred was in the parish of Cosby in a field which Nichols records as 'Guthlaxton Meadow' next to the Fosseway, a Roman road which forms the parish boundary. It is possible that *stan* refers to a Roman milestone.

Sparkenhoe Hundred

Sparchenhou 1130 (Leicestershire Survey)

This is either OE *spraeg* 'a shoot, twig' or OE *spearca* 'broom' + OE *hoh* 'a heel, a spur of land' 'A broom covered headland'. The meeting place lay one mile to the north of Peckleton on the Roman road from Mancetter to Leicester. The name of Shericles Farm here derives from OE *scirac* 'the Hundred or Shire oak'.

LEICESTERSHIRE SETTLEMENTS

Ab Kettleby   724230

Chetelbie 1086 (Domesday Book)

Ab this could be either the Middle English male personal name *Abba* or the OScand. male personal name *Abbi + Ketil* (OScand. personal name) + OScand. *by* 'farmstead, village'.

'The place associated with *Ketil*'.

The pre-fix **Ab**, which is not recorded until 1236, probably arose to distinguish this settlement from Eye Kettleby. It is unlikely that **Ab** dates from the period of the Viking settlement here.

Albert Village   303182

This settlement developed during the nineteenth century when the area was opened up for coal-mining. It was named after Prince Albert, Queen Victoria's consort.

Aldeby (deserted)   552987

Oldebi 1086 (Domesday Book)

OE *ald* 'old' + OScand. *by* 'farmstead, village'.

'The old settlement'.

This settlement may have already been in decline by the time of Domesday. The ruins of the church of St John and the deserted village can be seen by the river Soar.

Alderman's Haw   503145

Aldermanneshaga 1174 (Nichols)

OE *aldormann* 'a nobleman of the highest rank' + OE *haga* 'hedge or enclosure'.

'The nobleman's enclosure'.

This would have been a piece of land enclosed from the waste on the edge of Charnwood. The identity of this 'nobleman' is not known.

Allexton   818004
Adelachestone 1086 (Domesday Book)
Aethellac (OE personal name) + OE ***tun*** 'farmstead, village, small estate'.
'The place associated with ***Eadlac***'.

Alton (deserted)   390148
Heletone 1086 (Domesday Book)
OE ***ald*** 'old' + OE ***tun*** 'farmstead, village, small estate'.
'The old village'.
This probably means the old village in contrast with a new one.

Ambion (deserted)   400003
Anebem 1261 (Curia Regis Rolls)
OE ***an*** 'one, single' or ***ana*** 'lonely' + OE 'beam', 'a tree, a piece of heavy timber'.
This probably means 'a solitary tree'. Ambion disappeared from the records before the end of the fourteenth century. It was on Ambion Hill that Richard III is reputed to have harangued his army before the Battle of Bosworth Field, which actually took place on Redemore Plain.

Anderchurch (deserted)   392222
Andreskirka 1144 (Dugdale)
OE ***Andreas*** 'Andrew' + OE ***cirice*** 'church'.
'St Andrew's church'. This settlement lay between Bredon-on-the-Hill and Staunton Harold.

Anstey   549085
Anstige 1086 (Domesday Book)
OE ***anstiga*** 'a narrow footpath, often one that winds up a hill'.
'The settlement along the narrow footpath which winds up a hill'. This path probably originated as a track through woodland.

Appleby Magna   316099
Aeppelby 1002 (Anglo-Saxon will)

Aplebi 1086 (Domesday Book)
OE *aeppel* 'apple-tree' + OScand. *by* 'farmstead, village'.
'The settlement where apple-trees grow'.
Magna occurs for the first time in the early sixteenth century.

Appleby Parva (see Appleby Magna) 309088
'Parva' first appears in the late thirteenth century.

Arnesby 617923
Erendesberi 1086 (Domesday Book)
Iarund (OScand. male personal name) + OScand. *by* 'farmstead, village'.
'The settlement associated with *Iarund*'.

Asfordby 705192
Esseberie 1086 (Domesday Book)
Asfrothr (OScand. male personal name) + OScand. *by* 'farmstead, village'.
'The settlement associated with *Asfrothr*'.

Ashby-de-la-Zouch 355166
Ascebi 1086 (Domesday Book)
OE *aesc* 'ash tree(s) + OScand. *by* 'farmstead, village'.
'The settlement at or by the ash trees'. The manor was held by *Alanus la Zouche* in the early thirteenth century.

Ashby Folville 308120
Ascbi 1086 (Domesday Book)
OE *aesc* 'ash tree(s)' + OScand. *by* 'farmstead, village'.
'The settlement at or by the ash-trees'. *Fulco de Folevilla* held this manor in the early twelfth century.

Ashby Magna 563906
Essebi 1086 (Domesday Book)
OE *aesc* 'ash tree' + OScand. *by* 'farmstead, village'.
'The settlement at or by the ash trees'. 'Magna' was added in the early years of the thirteenth century to distinguish it from Ashby Parva. The two Ashbys would once have formed one land unit.

Ashby Parva 527884
(see Ashby Magna)

Ashby Woulds

This place, which was once part of the parish of Ashby-de-la-Zouch, was recognised as a separate entity in 1894.

Aston Flamville   463927

Erston 1209 (Episcopal Registers)

OE *east* 'east' + OE *tun* 'farmstead or village'.

'The east or eastern settlement', usually one lying to the east of a more important place. In 1100 this manor was held by Robert *de Flamville*, one of William the Conqueror's men.

Atterton (deserted)   353983

Altreton 1173 (Dugdale)

Aethelraed (OE male personal name) + OE *tun* 'farmstead, village, small estate'.

'The settlement associated with *Aethelraed*'.

Aylestone   573008

Ailestone 1086 (Domesday Book)

Aegel (OE male personal name) + OE *tun* 'farmstead, village, small estate'.

'The settlement associated with *Aegel*'.

Baggrave (deserted)   697088

Badegraue 1086 (Domesday Book)

Babba (OE male personal name) + OE *graf* 'grove, copse, thicket'.

'The grove associated with *Babba*'.

Bagworth   448080

Bageworde 1086 (Domesday Book)

Bacga (OE male personal name) + OE *worth* 'farmstead, enclosure'.

'The settlement associated with *Bacga*'.

Bardon   445128

Berdon 1240 (Feet of Fines)

OE *beorg* 'a hill, a mound, a burial mound' + OE **dun** 'a hill, an expanse of upland pasture'.

'The hill which has ancient burial mounds on it'. This hill has been quarried for granite leaving no trace of the prehistoric burial mounds to which it owes its name.

Barkby 637089

Barcheberie 1086 (Domesday Book)

Barki (OScand. male personal name) + OScand. *by* 'farmstead, village'.

'The settlement associated with *Barki*'.

Barkby Thorpe 636090

Torp 1199 (Curia Regis Rolls)

OScand. *thorp* 'outlying, dependent farmstead, hamlet'.

This village was settled from Barkby.

Barkestone 780349

Barchestone 1086 (Domesday Book)

Barkr (OScand. male personal name) + OE *tun* 'farmstead, village, small estate'.

'The place associated with *Barkr*'. This name comprises a Viking personal name with the OE *tun*, indicating that there had been a previous settlement here which had been taken over by Viking incomers.

Barlestone 428058

Beruluestone 1086 (Domesday Book)

Berwulf (OE male personal name) + OE *tun* 'farmstead, village, small estate'.

'The settlement associated with *Berwulf*'.

Barrow-upon-Soar 577176

Barhou 1086 (Domesday Book)

OE *bearu* 'wood, grove'.

'The wood or grove'. This name later became attached to a settlement'. *'Super Sore'* was first recorded in 1294.

Barsby 698113

Barnesbi 1086 (Domesday Book)

Barn (OScand. male personal name) + OScand. *by* 'farmstead, village'.

'The settlement associated with *Barn*'. This is a wholly Viking name in an area of the county where the names reveal dense Viking settlement.

Barton-in-the-Beans 395065

Bartone 1086 (Domesday Book)

OE **bere** 'barley' + OE **tun** 'farmstead, village, small estate'.

'The barley or corn farm'.

'In the Beans' was added in in 1251. The fertile soil here is well suited to growing arable crops such as barley or beans.

Barwell 445969

Barwalle 1043 (*Diplomatarium Anglicum*)

Barewelle 1086 (Domesday Book)

OE **bar** 'boar' + OE **wella** 'spring or stream'.

'The boar stream'.

Beaumanor 538158

Beumaner 1265 (Calendar of Charter Rolls)

OFr. **beau** 'beautiful' + OFr. **maner** 'manor, estate'

'Beautiful estate'.

Beaumont Leys 564075

OFr. **Beaumont lese** 1502

OFr. 'beautiful hill' + OE **leys** 'meadow'.

'The meadows near the beautiful hill'.

Beeby 664083

Bebi 1086 (Domesday Book)

OE **beo** 'bee' + OScand. **by** 'farmstead, village'.

'The settlement where bees are kept'. This is possibly a renaming of an already existing settlement with the Viking word **by** as a suffix to the OE **beo**.

Belgrave (Leicester) 593072

Merdegrave 1086 (Domesday Book)

OE **mearth** 'marten' + **graf** 'grove, copse, thicket'.

'The grove where martens proliferated'. This is one of the most interesting names in the county in that it appears to have been deliberately changed. By 1199 Merdegrave had become **Belagrave** 'beautiful grove'. The OE **mearth** had become associated with the OFr **merde** 'excrement' with the name clearly becoming unacceptable. This must have been a deliberate decision and not an evolution of the name as **mearth** would not have evolved naturally into **bel**.

Belton   448209

Beltone 1222 (*Rotuli* of Hugh de Welles)

OE *bel* 'space, interval, beacon' + OE *tun* 'farmstead, village, small estate'.

This could mean either 'a settlement in a clearing in woodland' or a place where a signalling beacon stood. At the time of Domesday this would have been a heavily wooded area. The village lies on land which rises slightly between two streams making the beacon interpretation a distinct possibility especially as Belton lies near to the county boundary as does Belton in Rutland.

Belvoir   818339

Belveder 1130 (Pipe Rolls)

OFr. *bel* 'beautiful' + OFr. *vedir* 'view'.

'Beautiful view'.

Belvoir Castle stands on a promontory of the wolds with extensive views across the Vale of Belvoir. The first castle on this site was founded soon after the Norman Conquest by Robert *de Todeni*, William's standard bearer.

Bescaby (deserted)   823263

Berthaldebia 1194 (Pipe Rolls)

Berg-Skald (OScand. male personal name) + OScand. *by* 'farmstead, village'.

'The settlement associated with *Berg-Skald*'.

Billesdon   719027

Billesdone 1086 (Domesday Book)

OE *Bill* (male personal name, which is a short form of *Bilheard)* + OE *dun* 'hill, an expanse of open hill country'.

'An expanse of high land associated with *Bill*'.

Given the dominance of this hill in the landscape for miles around it is reasonable to assume that this place would have been significant in the early Anglo-Saxon settlement of the area.

Bilstone   363054

Bildestone 1086 (Domesday Book)

Bild (OScand. male personal name) + OE *tun* 'farmstead, village, small estate'.

'The settlement associated with *Bild*. This place is a renaming of an already existing Anglo-Saxon settlement by the Viking incomer, *Bild*.

Birstall   597088
Burstelle 1086 (Domesday Book)
OE **burh** 'fortified place' + OE **stall** 'a place, a place which is the site of a building, a stall for cattle'.
This could mean either 'a fortified place, a disused fortified place, or cattle stall(s) at the disused fortified place'.

Bittesby (deserted)   500860
Bichesbie 1086 (Domesday Book)
Byttel (OE male personal name) + Oscan. **by** 'farmstead or village'.
'The settlement associated with **Byttel**'. This name combines an Anglo-Saxon personal name with a Viking habitative element an indication that this might be a renaming of an existing Anglo-Saxon settlement.

Bitteswell   537858
Betmeswelle 1086 (Domesday Book)
OE **bytme** 'head of a valley, valley' + OE **wella** 'spring or stream'.
'The spring or stream at the head of a valley'.

Blaby   570979
Bladi 1086 (Domesday Book)
Blar (OScand. male personal name) + Oscand. **by** 'farmstead or village'.
'The settlement associated with **Blar**'.

Blackfordby   329180
Blakefordebi 1130 (Leicestershire Survey)
OE **blaec** 'black' + OE **ford** 'ford + OScand. **by** 'farmstead, village'
'The settlement at the black ford'. This village lies in an area of coal deposits which must have coloured the stream bed.

Blaston   805965
Bladestone 1086 (Domesday Book)
Bleath (OE male personal name) + OE **tun** 'farmstead or village'.
'The settlement associated with **Bleath**'.

Boothorpe   320175
Bortrod 1130 (Leicestershire Survey)
Bo (OScand. male personal name) + OScand. **thorp** 'outlying, dependent farmstead or hamlet'.

'The settlement associated with **Bo**'.

Botcheston   843049

Borchardeston 1286 (Feudal Aids)

Bochard (OFr. male personal name) + OE **tun** 'farmstead, village, small estate'.

'The settlement associated with **Bochard**'. French names are rarely found in combination with the OE **tun**.

Bottesford   807392

Botesford 1086 (Domesday Book)

OE **botl** 'building, dwelling' + OE **ford** 'ford'.

'The ford at the building/dwelling'.

Botl is an interesting element, the literal meaning is 'a building' but in some parts of the country it refers to a significant dwelling, even a hall or palace. When this name was coined there would have been buildings everywhere so there must have been something remarkable about this particular building. Bottesford was an important central place in the Anglo-Saxon period so the suggestion that **botl** may be a 'palace' is not unreasonable.

Bradgate (deserted)   547108

Bradegate 1238 (Calendar of Close Rolls)

OE **brad** 'broad, wide, spacious' + OE **geat** 'an opening, a gap in the hills'.

'A broad gap in the hills'. Bradgate House was built in the gap. The settlement of Bradgate seems to have been located on a lane which running from Cropston westward to Bradgate Park. Enclosures and stone buildings could still be seen in 1943.

Bradley   816937

Bradele 1254 (Valuation of Norwich)

OE **brad** 'broad, wide, spacious' + OE **leah** 'wood or woodland clearing'.

'The wide clearing in the woodland'.

Branston   810295

Brantestone 1086 (Domesday Book)

Brant (OE male personal name) + OE **tun** 'farmstead, village, small estate'.

'The settlement associated with **Brant**'.

Braunstone (Leicester)   555028

Brandestone 1086 (Domesday Book)

Brant (OE male personal name) + OE *tun* 'farmstead, village, small estate'.

'The settlement associated with *Brant*'.

Brascote (deserted)   443025

Brocardescote 1086 (Domesday Book)

Brocheard (OE male personal name) + OE *cot* 'cottage, shelter'.

'The cottage associated with *Brocheard*'.

Breedon-on-the-Hill   405230

Briudun c.730 (Bede)

Celtic *briga* 'hill' + OE *dun* 'a hill, an expanse of open hill country'.

'A hill'. 'On the Hill' began to be added to the name in the seventeenth century by which time the original meaning of the elements *briga* and *dun* had long been forgotten. This hill, which is prominent in the landscape, has been an important settled site for more than 2,000 years. An Iron Age hill-fort was built here within the boundaries of which there was an important Anglo-Saxon monastery. The present church contains some of the most outstanding Anglo-Saxon stone carvings in England which would almost certainly have come from the earlier church of the monastery. It is surprising, given its earlier importance, that Breedon was not listed in Domesday Book.

Brentingby (deserted)   784198

Brantingbia 1130 (Leicestershire Survey)

Brant or *Brent* (OE male personal name) + OScand. *by* 'farmstead, village'.

'The settlement associated with *Brant/Brent*'. This place with an Anglo-Saxon personal name combined with *by* would seem to indicate that an already existing settlement had been taken over by incoming Vikings.

Bringhurst   842922

Brunungehyrst 1188 (Charter Rolls)

Bryni (OE male personal name) + OE *hyrst* 'hillock, copse, a wooded eminence'.

'The wooded eminence of *Bryni's* people'. In the early period of Anglo-Saxon settlement place-names often indicate 'folk' groups, whether these were kin groups or whether they were groups of people from the same place in the homeland is not known.

Bromkinsthorpe (deserted)   589042
Bruneskinnestorp 1086 (Domesday Book)
Bruncyng (OE male personal name) + OScand. *thorp* 'outlying, dependent farmstead, hamlet'.
'The settlement associated with *Bruncyng'*. This settlement lay just outside Leicester near Westcotes.

Brooksby (deserted)   670160
Brochesbi 1086 (Domesday Book)
Brok (OScand. male personal name) + OScan. *by* 'farmstead, village'.
'The settlement associated with *Brok'*.

Broughton Astley   526927
Broctone 1086 (Domesday Book)
OE *broc* + OE *tun* 'farmstead or village'.
'Settlement sited on a brook'. Astley derives from Thomas *de Estle* who held the manor in 1220.

Bruntingthorpe   604898
Brandinestor 1086 (Domesday Book)
Brenting (OE male personal name) + OScand. *thorp* 'outlying dependent settlement'.
'The settlement associated with *Brenting'*.

Buckminster   880229
Bucheminstre 1086 (Domesday Book)
Bucca (OE male personal name) + OE *mynster* 'minster'. The church here, presumably founded by *Bucca*, would have been an Anglo-Saxon missionary centre, supporting dependent chapelries in settlements in the surrounding countryside.

Burbage   443925
Burbece 1086 (Domesday Book)
OE *burh* 'fortified place' + OE *bece* 'brook, valley'.
'The brook or valley by the fortified place'.

Burrough-on-the-Hill   758108
Burg 1086 (Domesday Book)
OE *burh* 'fortress, fortified place'.

The impressive Iron Age hill-fort which lies to the north of the village would have been an important gathering place of the Iron Age tribe, the Corieltauvi, whose territory covered Leicestershire and much of Lincolnshire until the coming of the Romans in AD 43. Excavations have produced evidence of Romano-British occupation although the fort would almost certainly have ceased to serve its previous function under Roman rule.

Burton Lazars   769169

Burtone 1086 (Domesday Book)

OE *burh* 'a fortified place' + OE *tun* 'farmstead, village, small estate'. 'The settlement at the fortified place'. Roger *de Mowbray* founded a leper hospital here in 1138.

(See Burton Overy for a detailed explanation of Burton names)

Burton-on-the-Wolds   590211

Bortone 1086 (Domesday Book)

OE *burh* 'fortified place' + OE *tun* 'farmstead, village, small estate'. 'The settlement at the fortified place'.

The suffix 'on the Wolds' is first recorded in 1301 as '*super Waldas*', this usage would have arisen to distinguish this place from Burton Overy and Burton Lazars.

(See Burton Overy.)

Burton Overy   677980

Burtone 1086 (Domesday Book)

OE *burh* 'fortified place' + OE *tun* 'farmstead, village, small estate'.

The compound name '*burh-tun*' denotes an area that has been specifically established as a fortified place. The three Burtons in Leicestershire share the same meaning. Most Burtons are found in those areas of England where there were serious threats in the ninth century from Viking invasions; *burh-tuns* appear to have been deliberately established to counteract that threat. Clearly they did not succeed.

Bushby   653043

Bucebi 1175

Butr (OScand. male personal name) + OScand. *by* 'farmstead, village'.

The settlement associated with *Butr*.

Cadeby 426024

Catebi 1086 (Domesday Book)

Kati (OScand. male personal name) + OScand. *by* 'farmstead, village'.

'The settlement associated with *Kati*'.

Carlton 397050

Karlintone 1209 (Episcopal Registers)

OE *ceorl* 'a churl, a freeman, a peasant' + OE *tun* 'farmstead, village, small estate'.

The settlement of the free peasants here would have been part of a large Anglo-Saxon estate centred on Market Bosworth.

(See Carlton Curlieu)

Carlton Curlieu 694973

Carlintone 1086 (Domesday Book)

OE *ceorl* ' a churl, a freeman, a peasant' + OE *tun* 'farmstead, village, small estate'.

The most probable meaning of this name is that of 'the enclosure of the free peasants'. Many Charltons are found close to important central places (in this case Great Glen which was an Anglo-Saxon royal centre). The modern word churlish comes from *ceorl* and must reflect the fact that being free produced a surly attitude.

Castle Donington 448275

Dunitone 1086 (Domesday Book)

Dunn (OE male personal name) + OE *inga-tun* 'farmstead, village, small estate'.

'The settlement associated with *Dunn*'. *Castel* 'castle' appears in the records for the first time in 1331.

Catthorpe 553781

Torp 1086 (Domesday Book)

Cat (OScand. male personal name) + OScand. *thorp* 'outlying, dependent farmstead or hamlet'.

'The farmstead or hamlet associated with *Cat*'. *Cat* appears in the records in 1232 as *Thorp le Cat*.

Chadwell 782246

Caldeuuelle 1086 (Domesday Book)

OE *cald* 'cold' + OE *wella* 'spring or stream'.

This is probably 'the cold spring'.

Charley   476142

Cernelega 1086 (Domesday Book)

Celtic ***carn*** 'a heap of stones, or a cairn', + OE ***leah*** 'wood, clearing in a wood'.

'A woodland clearing among/near rocks and stones. This name is more likely to refer to the rocky terrain here rather than to a specific cairn, and would almost certainly have applied to a wide area. The sixteenth century the traveller and antiquarian William Leland refers to 'the *foreste* of Charley', an earlier name for Charnwood Forest.

Charnwood   509148 (Beacon Hill)

Charnewode 1276 (*Rotuli Hundredorum*)

Celtic ***carn*** 'a heap of stones, a cairn' + OE ***wudu*** 'a wood'.

'A wood near a rocky landscape'.

Chilcote   285115

Cildecote 1086 (Domesday Book)

OE ***cild*** 'child', ***cilda*** 'children' + OE ***cot*** 'cottage'. This is literally the children's cottage but the child(ren) in this case most probably refers to a noble born youth or youths. This place could well have been a farm or small estate established for the younger son(s) of a landed/noble family.

Church Langton   723923

Langetone 1086 (Domesday Book)

OE ***lang*** 'long' + OE t***un*** 'farmstead, village, small estate'.

'The long settlement'. 'Church' is first recorded in 1316. Church Langton was the central place of a large land unit which included all the Langtons. The church, which stands on a hill, can be clearly seen from all the component settlements of the estate.

Claybrooke Magna   493888

claeg broc 962 (*Cartularium Saxonicum*)

Claibroc 1086 (Domesday Book)

OE ***broc*** 'brook' + OE ***claeg*** 'clay'.

'The brook with the clay bed, the clayey brook'.

'Magna' first appears in the record in 1261 as also does Parva.

Claybrooke Parva   492880
(See Claybrook Magna)

Coalville   428141
Coalville 1838 (County Rate Return)
Nineteenth century name for the town which grew up in the centre
of the Leicesteshire coalfield. Coalville lies within the ancient parish
of Snibston.

Cold Newton (deserted)   716065
Niweton 1086 (Domesday Book)
OE *niwe* 'new' + OE *tun* 'farmstead, village, small estate'.
'The new settlement'. OE *calde* 'cold' appears in the record for the
first time in1279. Seemingly, from early in its existence, this was a
difficult place to farm.

Cold Overton   811101
Ovretone 1086 (Domesday Book)
OE *ofer* 'ridge' + OE *tun* 'farmstead, village, small estate'.
'The settlement on the ridge'. By 1130 the affix OE *cald* 'cold, bleak,
exposed' is being used.

Coleorton   414175
Ovretone 1086 (Domesday Book)
OE *ofer* 'bank, ridge' + OE *tun* 'farmstead, village, small estate'.
'The settlement at or near the ridge'. The affix OE *col* 'coal' appears
for the first time in 1443 which must be when the exploitation of
coal in this area began.

Congerstone   368055
Cuningestone 1086 (Domesday Book)
OE *cyning* 'king + OE *tun* 'farmstead, village, small estate'.
'The settlement with royal associations or serving a royal function'.
This place may once have been part of an extensive Anglo-Saxon
royal estate centred on Market Bosworth, its exact function is not
known.

Copt Oak   482129
le Coppudhok 1230 (Leicester Museum MS.)
OE *copped* 'pollarded' + OE *ac* 'an oak tree'. This must have referred

39

to a distinctive pollarded oak tree, one perhaps which lay on a boundary.

Cosby 548955

Cossebi 1086 (Domesday Book)

This is either ***Cossa*** or ***Kofsi*** (OScand. male personal name) + OScand. ***by*** 'farmstead or village'.

'The settlement associated with either ***Cossa*** or ***Kofsi***'.

Cossington 606135

Cosintone 1086 (Domesday Book)

Cossa (OE male personal name) + OE ***tun*** 'farmstead, village, small estate'.

'The settlement associated with ***Cossa***'.

Coston 849222

Castone 1086 (Domesday Book)

Katr (OScand. male personal name) + OE ***tun*** 'farmstead, village, small estate'.

'The settlement associated with ***Katr***'. The OScand. personal name with the OE suffix ***tun*** indicates that there was an already existing Anglo-Saxon settlement here before the coming of the Vikings in the second half of the ninth century.

Cotes 556208

Cotes 1200 (Danelaw Charters)

OE ***cotts***.

'Cottages' or 'shelters for animals' (usually sheep).

Cotesbach 537824

Cotesbece 1086 (Domesday Book)

Cott (OE male personal name) + OE ***bece*** 'stream, valley'.

'The valley with a stream associated with ***Cott***'.

Cotes de Val (deserted) 553887

Toniscote 1086 (Domesday Book)

Tone (OE male personal name) + ***cot*** 'cottage, shelter'.

By 1194 this name is just ***Cotes***, de Val is unclear, there was no local family of that name although in nearby Warwickshire there was a *d'Eyvill* family who held lands there in the thirteenth century, it is possible that they also had holdings here.

Countesthorpe   586955

Torp 1156 (Calendar of Charter Rolls)

OScand. *thorp* 'outlying, dependent farmstead or hamlet'.

Cuntasse appears in the record for the first time in 1276.

'The countess's outlying settlement'. The 'countess' may have been connected with Simon de Montfort, Earl of Leicester, who held lands here in 1265.

Cranoe   761952

Craweo 1086 (Domesday Book)

OE *crawena* 'crows' + OE *hoh* 'a heel or a spur of land'.

'The headland frequented by crows'. Cranoe lies on a small spur of land in the valley of the River Welland.

Croft   511960

Craeft 836

Crebre 1086 (Domesday Book)

OE *cræft* 'machine, engine'.

This name probably refers to a water mill. In the ninth century Croft was an important Anglo-Saxon royal centre. The name must refer to something which was, in its time, outstandingly remarkable. The hill here is a prominent feature in a flat landscape.

Cropston   554109

Cropeston 1130 (Leicestershire Survey)

This could either be *Cropp* (OE male personal name) *or Kroppr* (OScand. male personal name) + *tun* 'farmstead, village, or small estate'.

'The settlement associated with *Croppa/Kroppr'*.

Croxton Kerrial   835924

Crontone 1086 (Domesday Book)

Croc (OScand. male personal name) + OE *tun* 'farmstead, village or small estate'.

'The settlement associated with *Croc'*.

The Scandinavian personal name combined with the OE *tun* indicates that this was almost certainly an Anglo-Saxon settlement which was taken over by a Viking incomer. This manor was granted to *Bertram de Cryoil* in 1242, the name derives from Criel in northern France.

Dadlington   405980
Dadelintone 1216 (*Rotuli Litterarum Clausarum*)
Dadela (OE male personal name) + OE *tun* 'farmstead, village, small estate'.
'The settlement associated with *Dadela*'.

Desford   478034
Diresford /Deresford 1086 (Domesday Book)
Deor (OE male personal name) + **OE *ford*** 'ford'.
'The ford associated with *Deor*'. *Deor* probably held the land where the ford lay.

Diseworth   854245
Diwort 1086 (Domesday Book)
Digoth (OE male personal name) + OE *worth* 'enclosure, village'.
The settlement associated with *Digoth*.

Dishley (deserted)   513212
Dislea 1086 (Domesday Book)
Digoth's (OE male personal name) + OE *leah* 'wood, clearing in a wood'. Dishley, which is only four miles away, also records a *Digoth*; there may have been a connection between the two settlements.

Donington–le–Heath   420125
Duntone 1086 (Domesday Book)
Dunn (OE personal name) + OE *tun* 'farmstead, village, small estate'
'The settlement associated with *Dunn*'. '*Super le heth*' appears in the records in 1449.

Donisthorpe   315140
Durandestorp 1086 (Domesday Book)
Durand (OE male personal name) + OScand. *thorp* 'outlying or dependent settlement'.
'The outlying settlement associated with *Durand*'.

Drayton   831922
Draiton 1163 (Pipe Rolls)
OE *draeg* 'a drag, a portage, a slipway' + OE *tun* 'farmstead, village, small estate'.
'The place where it is necessary to drag or haul'.

Drayton is at the foot of a hill that rises steeply up to Nevill Holt. Holt is OE for wood or thicket so the **draeg** might refer to the where felled trees were hauled down to the river.

Dunton Bassett   547907

Donitone 1086 (Domesday Book)

OE **dun** 'a hill, an upland expanse' + OE **tun** 'farmstead, village, small estate'.

'The settlement on the hill'. *Ralf Basset* held the manor in 1166.

Earl Shilton   470980

Sceltone 1086 (Domesday Book)

OE **scelf** 'a bank, ledge, shelf' + OE **tun** 'farmstead, village, small estate'.

'The settlement on the bank or shelf'. This shelf can be seen clearly on the A47 as it leaves the town on the Leicester side. 'Earl' refers to Edmund Plantagenet, Earl of Leicester, son of Henry III.

East Goscote   645136

Building development began here in the 1960s and was named after East Goscote Hundred in which it lies.

Easthorpe   810386

Esttorp 1240 (Duke of Rutland's Mss.)

OE **east** 'east, eastern' + OE **thorp** 'outlying/dependent farmstead'.

'The eastern outlying settlement'. This settlement lies east of Bottesford.

East Langton   726926

Lagintone, Langetone 1086 (Domesday Book)

OE **lang** 'long' + OE **tun** 'farmstead, village, small estate'.

'The long settlement'. 'East' refers to the settlement lying east of Church Langton which was the central place of all the Langtons when they were one large land unit.

East Norton   784005

Nortone 1086 (Domesday Book)

OE **north** 'north' + OE **tun** 'farmstead, village, small estate'.

'The north(ern) settlement'. North of the important royal Anglo-Saxon centre of Great Glen. 'East' was added later to distinguish this place from Kings Norton.

Eastwell   777286
Estwelle 1086 (Domesday Book)
OE *east* 'east + OE *wella* 'spring or stream'.
'The east stream'. The River Devon flows eastwards from here.

Eaton   798290
Aitona 1130 (Leicestershire Survey)
OE *eg* 'an island, dry land amidst wet, fenny land + OE *tun* 'farmstead, village, small estate'.
'The area of dry land in a fenny area'.

Edmondthorpe   858177
Edmerestorp 1086 (Domesday Book)
Eadmaer (OE male personal name) + OScand. *thorp* 'outlying, dependent, secondary farmstead, hamlet'.
'The settlement associated with *Eadmaer*'.

Ellistown   430112
This new coal-mining town was named after John Ellis who opened a colliery here in 1875.

Elmesthorpe   460965
Aylmerestorp 1207 (Curia Regis Rolls)
Aethelmaer (OE/OScand. male personal name) OScand. *thorp* 'outling, dependent farmstead, hamlet'.
'The outlying settlement associated with *Aethlmaer*'. This personal name could be in origin either Anglo-Saxon or Viking. It is not recorded until the 13th century making it difficult to be sure. The site of this settlement has been subsumed into Earl Shilton.

Enderby   537994
Andretesbie 1086 (Domesday Book)
Eindridi (OScand. male personal name) + OScand. *by* 'farmstead, village'.
'The settlement associated with *Eindridi*'.

Evington   627027
Avintone 1086 (Domesday Book)
Eafa (OE male personal name) + OE *tun* 'farmstead, village, small estate'.
'The place associated with *Eafa*'.

Eye Kettleby (deserted) 734167
Chitebie 1086 (Domesday Book)
Ketil (OScand. personal name) + OScand. *by* 'farmstead, village'.
'The place associated with *Ketil'*. Eye refers to the River Eye which flows nearby and distinguishes this place from Ab Kettleby which lies four miles to the north-west.

Far Coton 387021
Cotes 1200 Curia Regis Rolls
OE *cot* 'cottage(s) huts.
'The cottage(s)'. 'Far' first appears on a map of 1785.

Fenny Drayton 351970
Draitone 1086 (Domesday Book)
OE *draeg* 'drag, a portage, a slipway' + OE *tun* 'farmstead, village, small estate'.
'A place where it is necessary to drag or haul'. 'Fenny' was added in the fourteenth century, this OE *fennig* 'fenny, marshy'. This village lies on ground that was once marshy. The fenny terrain was probably the reason for the '*draeg'*.

Fleckney 648935
Flecchenie 1086 (Domesday Book)
Flecca (OE male personal name) + OE *eg* 'an island, piece of dry land in a fen'.
'The dry land in fenny land associated with *Flecca'*.

Foston (deserted) 604950
Fostone 1086 (Domesday Book)
Fotyr (OScand. male personal name) + OE *tun* 'farmstead, village, small estate'.
'The settlement associated with *Fotyr'*. This name combines a Viking personal name with the OE *tun*, an indication that an earlier settlement may have been taken over by a Viking incomer.

Foxton 700900
Foxestone 1086 (Domesday Book)
OE fox '*fox'* + OE *tun* 'farmstead, village, small estate'.
'The place where foxes abounded'.

Freeby   806201
Fredebi 1086 (Domesday Book)
Fraethi (OScand. male personal name) + OScand. *by* 'farmstead, village'.
'The settlement associated with *Fraethi*'.

Frisby-by-Gaulby (deserted)   704020
Frisebie 1086 (Domesday Book)
OE *Frisa* 'native of Friesland or the Frisian Isles' + OScand. *by* 'farmstead or village'.
'The settlement of the Frisians'. Frisby on the Wreake also refers to Frisians who must have stood out as a distinct group'. The OScand. *by* indicates that the name dates from the Viking period although the Frisians could have been settled here earlier with the incoming Vikings adding the suffix *by*.

Frisby-on-the-Wreake   696178
Frisebie 1086 (Domesday Book)
(see Frisby by Gaulby)

Frolesworth   504907
Frellesworde 1086 (Domesday Book)
Freothulaf (OE male personal name) + OE *worth* 'enclosure, farmstead, village'.
'The settlement associated with *Freothulaf*'.

Gaddesby   688131
Gadesbie 1086 (Domesday Book)
Gaddr (OScand. male personal name) + OScand. *by* 'farmstead, village'.
'The settlement associated with *Gaddr*'.

Garendon (deserted)   502199
Geroldone 1130 (Leicestershire Survey)
Gaerwald (OE male personal name) + OE *dun* 'a hill, an expanse of open hill country'.
'The hill associated with *Gaerwald*'. Garendon Abbey was founded here in 1132.

Garthorpe   833209
Garthorp 1130 (Leicestershire Survey)
This could be either OE *gara* 'a gore, a triangular-shaped plot of

land' or OScand. *gardr* 'an enclosure, yard, garden' (especially one near a house) + OScand. *thorp* 'outlying, dependent farmstead or hamlet'.

'A farmstead within an enclosed corner of land'. There is a V-shaped valley just to the north of Garthorpe.

Gaulby 695010

Galbi 1086 (Domesday Book)

OE *gall* 'gall, a barren or wet place' + OScand. *by* 'farmstead, village'.

'The settlement on barren soil'.

Gilmorton 572879

Mortone 1086 (Domesday Book)

OE *mor* 'moorland' + OE *tun* 'farmstead, village, small estate'.

'The settlement on moorland'. The OE *gylden* 'golden', appears for the first time in 1303 as *Gildenemorton* and probably refers to a yellow-coloured plant such as broom.

Glen Parva 566988

(see Great Glen)

Glenfield 546057

Clanefelde 1086 (Domesday Book)

OE *claene* 'clean' + OE *feld* 'tract of open country'.

'Tract of open country without thorns or overgrowth'. *Feld* later developed to mean land for pasture or cultivation such as the unenclosed open fields of much of medieval Leicestershire and Rutland. The modern sense of the word 'field', meaning 'an enclosed or fenced-in plot of land', seems to have first come into use in the fourteenth century when massive population decline after the various visitations of the Black Death led to large tracts of land being enclosed for pasture. The Parliamentary enclosure of arable land in the eighteenth and nineteenth centuries resulted in a limitation of meaning to field.

Glooston 751959

Glorestone 1086 (Domesday Book)

Glor (OE male personal name) + OE *tun* 'farmstead, village, small estate'.

'The settlement associated with *Glor*'.

Goadby   750988
Goutebi 1086 (Domesday Book)
Gauti (OScand. male personal name) + OScand. *by* 'farmstead, village'.
'The settlement associated with *Gauti*'.

Goadby Marwood   780264
Goutebi 1086 (Domesday Book)
Gauti (OScand. male personal name) + OScand. *by* 'farmstead, village'.
'The settlement associated with *Gauti*'.

Gopsall (deserted)   353064
Gopeshille 1086 (Domesday Book)
OE *gop* 'serf, slave' + OE *hyll* 'hill'.
'The hill associated with the slaves'.
The Anglo-Saxons were a slave-owning society, Domesday Book records many slaves attached to some manors. This settlement may have had a specialist function which is impossible to discover.

Great Bowden   746889
Bugedone 1086 (Domesday Book)
Bucga (OE female personal name) + OE *dun* 'a hill, a flat-topped hill, an open upland expanse'.
'The open upland country associated with *Bucga*'. OE *dun* is the modern word 'downs', the 'down' of Great Bowden can be clearly seen on the road that leads into Bowden from the A6 to the north of Market Harborough, it is a typical *dun* with a fairly level extensive summit. The name *Bucga* is of particular interest as it is a woman's name used in a place-name, one of only two in Leicestershire; the other one is Witherley. At the time of Domesday Bowden was a major Anglo-Saxon royal centre held by King Edward the Confessor.

Great Dalby   745145
Dalbi 1086 (Domesday Book)
OScand. *dalr* 'valley' + OScand. *by* 'farmstead, village'.
The settlement in the valley. 'Great' distinguishes this village from Little Dalby. These two settlements would initially have been one land unit.

Great Easton 848931

Estone 1086 (Domesday Book)
OE *east* 'east, eastern' + OE *tun* 'farmstead, village, small estate'.
'The settlement lying to the east of a more important place', in this case Medbourne.

Great Glen 655980

Aet Glenne 849 (Anglo-Saxon (royal) Charter)
Glen 1086 (Domesday Book)
Celtic *glennos, glen, glin* 'a glen, a valley'.
Glen is the old name of the river Sence used as a settlement name. *Glen* usually describes a wide valley which is low-lying in relation to the surrounding area and which is subject to flooding. Between Great Glen and Wistow the river meanders across flood plains and the records of Wistow church reveal that the site has always been subject to flooding.

In the ninth century Great Glen was the centre of an Anglo-Saxon royal estate that stretched from the source of the river near Billesdon to at least Newton Harcourt and Fleckney. The name of Glen Parva, sited where the River Glen/Sence joins the River Soar, would suggest that the Anglo-Saxon royal estate stretched along the whole length of the river; unfortunately there is no evidence to support this.

Great Stretton (deserted) 657005

Stratone 1086 (Domesday Book)
OE *straet* 'street, Roman road' + OE *tun* 'farmstead, village, small estate'.
'The settlement on the Roman road'. The Roman road here is the Gartree road which ran from Leicester to Colchester. Places with the name Stretton are usually sited on or very near to Roman roads.

Griffydam 413189

Griffydam 1676 (Leicestershire Wills)
OScand. *gryfja* 'a cleft, a deep valley' + OE *damme* or OScand. *dammr* 'dam'.
'A dam or a bank across a deep valley'.

Grimston 685219

Grymestone 1086 (Domesday Book)
Grim (OScand. male personal name) + OE *tun* 'farmstead, village,

small estate'.

'The settlement associated with **Grim'**. This name indicates that a Viking has taken over an already existing Anglo-Saxon settlement. **Grim** is the name that the Vikings gave to the god *Othin/Odin* when he appeared in disguise. It is also a common male personal name.

Groby   523076

Grobi 1086 (Domesday Book)

OE **grof** 'pit' + OScand. **by** 'farmstead or village'.

The 'pit' of this name must refer to Groby Pool, a large expanse of water which lies in a deep hollow and into which several streams flow'.

Gumley   683900

Godmundeslaech AD749 (Anglo-Saxon (royal) Charter)

Gotmundeslea 1086 (Domesday Book)

Godmund (OE male personal name) + **leah** 'woodland, woodland clearing'.

'The woodland/woodland clearing associated with **Godmund**'.

Gumley was an Anglo-Saxon royal estate in the eighth century, when both King Aethelbald and King Offa held meetings of their royal councils. The function of this estate at Gumley was probably, given the nature of the countryside, a royal hunting lodge. By the time of Domesday there was no hint of its former glory.

Hallaton   789967

Alctone 1086 (Domesday Book)

OE **halh** 'nook or corner of land, narrow secluded valley' + OE **tun** 'farmstead, village, small estate'.

'The settlement in a corner of land'.

Halstead (deserted)   750057

Elstede 1086 (Domesday Book)

OE **hald** 'shelter, refuge' + OE **stede** 'place, site, locality'.

'A place of shelter' probably for livestock.

Hamilton (deserted)   645075

Hamelton 1130 (Leicestershire Survey)

Hamela (OE personal name) + OE **tun** 'farmstead, village, small estate'.

'The settlement associated with **Hamela**'.

The earthworks of this deserted village can still be seen from a public footpath which crosses the site (see page 19).

Harby   747313

Herdebi 1086 (Domesday Book)

This is either OE **heorde** or OScand. **hjorth**. Both carry the same meaning of either a 'herd or a flock' + OScand. **by** 'farmstead, village'.

'The farmstead or village of the herdsmen'. This name might indicate that this was a specialist settlement, attached to a larger estate, where the herdsmen lived and folded the flocks when they were not out in the field.

Harston   839318

Herston 1086 (Domesday Book)

OE **har** 'grey' or a 'stone covered with lichen' + OE **stan** 'stone'.

This is probably 'the boundary stone'. There are other examples of **har** + **stan** which seem to indicate that the term carries a meaning of greater significance than a simply a 'grey stone'. Harston place-names are often occur on important boundaries. This place lies on the boundary between Leicestershire and Lincolnshire which here follows the line of a long-distance ancient trackway, used by the Romans, and known as the Sewstern Way

Hathern   502224

Avederne 1086 (Domesday Book)

OE **hagu-thyrne** 'white-thorn'.

'The white-thorn (hawthorn)'.

Heather   390108

Hadre 1086 (Domesday Book)

OE & OScand. **heithr** 'heather'.

'Heath, heathland' which stretched to Charnwood Forest and was close to Donington-le-Heath and Normanton-le-Heath. The village took its name from this expanse of heathland.

Hemington   455280

Aminton 1125 (Leicestershire Survey)

Hemma (OE male personal name) + OE **inga-tun** 'farmstead, village, small estate'.

'The settlement of **Hemma's** people'. This is a type of name which is associated with an early period of Anglo-Saxon settlement'.

Higham-on-the-Hill   385955
Hecham 1173 (Dugdale)
OE **heah** 'high' + OE **ham** 'farmstead, village'.
'The high settlement'. Although the OE element **ham** occurs early in the period of Anglo-Saxon settlement this place is not recorded in Domesday. It would have been in existence for at least 300 years by the time of Domesday and must have been listed under another place, probably Hinckley.

Hinckley   427938
Hinchelie 1086 (Domesday Book)
Hinca (OE male personal name) + OE **leah** 'woodland, woodland clearing'.
'The woodland clearing associated with **Hynca**'.

Hoby   669174
Hobie 1086 (Domesday Book)
OE **hoh** 'a heel, a hill spur' + OScand. **by** 'farmstead, village'.
'The settlement at or near the hill spur'.

Holwell   737237
Hollewelle 1086 (Domesday Book)
OE **hol** 'hollow', 'deep' + OE **wella** 'spring or stream'.
'A stream or spring in a deep hollow'.

Horninghold   807972
Horniwale 1086 (Domesday Book)
Horna (OE male personal name) + OE **ingas** + OE **wald** 'forest, woodland, high open ground'.
Horna + ingas is a folk name which indicates this is the **wald** associated with the people or followers of **Horna**. **Horn** might also refer to the shape of the valley here. If that is the case then the meaning of the name would be 'the folk who inhabit the horn-shaped valley'. Either of these meanings could be correct and certainly a 'horn' could describe the valley here. Folk names usually indicate early Anglo-Saxon settlement.

Hose   738294

Hoches 1086 (Domesday Book)

OE *hoh* 'a heel, a hill spur', the plural is *hohas*. This village lies in the Vale of Belvoir, above it lie the wolds along which there are several spurs.

Hoton   574226

Hohtone 1086 (Domesday Book)

OE *hoh* 'a heel, a hill-spur' + OE *tun* 'farmstead, village, small estate'. 'The settlement at or near the spur of land'. The *hoh* is not so pronounced here as at Hose (above) but nevertheless can still be seen.

Houghton-on-the-Hill   676033

Hohtone 1086 (Domesday Book)

OE *hoh* 'a heel a hill spur' + OE *tun* 'farmstead, village, small estate'. 'The settlement on the spur of the hill'.

Hugglescote   427127

Hukelescot 1227 (Charter Rolls)

Hucel (OE male personal name) + OE *cot* 'cottage'.
'The cottage(s) associated with *Hucel*'.

Humberstone   626060

Humerstane 1086 (Domesday Book)

Hunbeorht (OE male personal name) + OE *stan* 'stone'.

Hunbeorht's stone. This refers to an ancient standing stone, which, at the time the name was coined, must have stood on the land associated with *Hunberht* .

Huncote   516975

Hunecote 1086 (Domesday Book)

Huna (OE male personal name) + OE *cot* 'cottage, hut, shelter'.
'Cottage(s) associated with *Huna*'.

Hungarton   691073

Hungretone 1086 (Domesday Book)

OE *hungor* 'hungry' + OE *tun* 'farmstead, village, small estate'
'The settlement on barren soil'. The name of nearby Galby also refers to sterile soil and Cold Overton is not far away.

Husbands Bosworth   645845

Bareswerde 1086 (Domesday Book)

Bar (OE male personal name) + OE **worth** 'an enclosure, a farmstead'.

'The enclosure associated with **Bar**'. **Worth** as a place-name element is found in documents as early as the seventh century, it always denotes 'enclosure, fence, hedge'.

The prefix Husbands was added in the sixteenth century to distinguish it from Market Bosworth.

Ibstock   407105

Ibestoche 1086 (Domesday Book)

Ibba (OE male personal name) + OE **stoc** 'dairy farm'.

'The dairy farm associated with **Ibba**'.

Ilston-on-the-Hill   708993

Elvestone 1086 (Domesday Book)

Iolfr (OScand. male personal name) + OE **tun** 'farmstead, village, small estate'.

This Viking personal name with the Anglo-Saxon suffix **tun** indicates that an existing Saxon settlement was taken over by a Viking incomer who renamed it after himself.

Ingarsby (deserted)   684055

Inwaresbie 1086 (Domesday Book)

Ingvar (OScand. male personal name) + OScand. **by** 'farmstead, village'.

'The settlement associated with **Ingvar**'. The earthworks of the medieval village can clearly be seen here.

Isley Walton   426250

Waleton 1185 (*Records of the Templars in England*, BA Lees, 1935)

OE **walh** 'Briton, serf/slave' + **tun** 'farmstead, village, small estate'.

'The settlement of the 'natives, serfs, slaves'. For a detailed discussion of the element **walh** see Walcote. '*Isly*', which appears for the first time in 1327, is probably the name of a settlement both lost and unrecorded in its own right. Without any early spellings it is difficult to be sure about this name but the second element is probably OE **leah** 'woodland, woodland clearing' and **Isa** may have been an OE male personal name.

Kegworth   488267
Cacheworde 1086 (Domesday Book)
Caegga (OE male personal name) + OE **worth** 'enclosure, village'.
'The settlement associated with **Caegga**'.

Keyham   670066
Caiham 1086 (Domesday Book)
Caega (OE male personal name) + OE **ham** 'homestead, village'.
'The homestead or village associated with **Caega**'. Place-names which contain the habitative element **ham** date from early in the period of anglo-Saxon settlement'.

Keythorpe (deserted)   765955
Cheitorp 1086 (Domesday Book)
Keyia (OScand. male personal name) + OScand. **thorp** 'outlying dependent hamlet'.
'The settlement associated with **Keyia**'.

Kibworth Beauchamp   685935
Chiburde 1086 (Domesday Book)
Cybba (OE male personal name) + OE **worth** 'farmstead, enclosure'.
'The enclosed settlement associated with **Cybba**'. *Walter de Bellocampo* is recorded as holding this manor in 1130 Leicester Survey. The two Kibworths were once one land unit.

Kibworth Harcourt
(See Kibworth Beauchamp)
Ivo de Haruecurt held this manor in the twelfth century.

Kilby   623954
Cilebi 1086 (Domesday Book)
The element **by** is probably a Scandinavianised form of the OE **cildatun** which means literally 'the settlement of the children'. The 'children' would have been young men of noble birth. They may have actually lived on the estate or, more likely, the estate provided them with a livelihood. Neighbouring Great Glen was an important Anglo-Saxon royal centre and Kilby was probably once part of that royal estate.

Kilwardby (deserted)   355166
Culverteby 1130 (Leicestershire Survey)

Kilvert (OScand. personal name) + OScand. ***by*** 'farmstead, village'. 'The settlement associated with Kilvert'. This place lay in the west of of the parish of Ashby-de-la-Zouch.

Kimcote   586865

Chenemundescote 1086 (Domesday Book)
Cynemund (OE male personal name) + OE ***cot*** 'cottage, shelter' **'Cynemund's** cottage'.

Kings Norton   689005

Nortone 1086 (Domesday Book)
OE ***north*** + OE ***tun*** 'farmstead, village, small estate'.
'The north(ern) settlement'. These names which describe orientation usually stand in relation to a more important place to which they are often attached, in this case the Anglo-Saxon royal centre of Great Glen.

Kirby Bellars   717177

Cherchebi 1086 (Domesday Book)
OScand. compound word ***kirkju-by*** 'the church settlement'.
'The settlement with a church'. This is Viking name which would have been given to an Anglo-Saxon church and settlement that was already in existence when the Vikings came. There would have been an earlier Anglo-Saxon name for this place. This manor was held by *Hamo Beler* in 1166.

Kirby Muxloe   520046

Carbi 1086 (Domesday Book)
Kaerir (OScand. male personal name) + OScand. ***by*** 'farmstead, village'.
'The settlement associated with ***Kaerir***. 'Muxloe', which is a local surname appearing first in the seventeenth century was adopted to distinguish this place from Kirby Bellars.

Kirkby Mallory   546007

Cherchebi 1086 (Domesday Book)
OScand. compound word ***kirkju-by*** 'church-settlement'.
This is a Viking name for an existing Anglo-Saxon settlement with a church. This manor was held by *Ricardus Malore* in the early thirteenth century.

Knaptoft (deserted)   626895
Cnapetoft 1086 (Domesday Book)
Cnapa (OE male personal name) + OScand. *toft* 'plot of land'.
'*Cnapa's* plot of land'.

Knighton (Leicester)
Cnichtetone 1086 (Domesday Book)
OE *cniht* 'knight, nobleman, retainer' + OE *tun* 'farmstead, village, small estate'.
'The settlement of the knights or noble retainers'.

Knipton   826313
Gnipetone 1086 (Domesday Book)
OScand. *gnipa* 'steep hillside' + OE *tun* 'farmstead or village'.
'The farmstead or village below or near a steep hillside'. The Scandinavian word *gnipa* is an indication that Vikings were settled here.

Knossington   801086
Nossitone 1086 (Domesday Book)
The meaning of this name is unclear. It could be OE *cnoss* 'hill', OScand. *knauss* 'rounded hill, or the OE *Cnoss(a)* (male personal name) + OE *tun* 'farmstead, village, small estate'.
This is either 'the settlement at or near the round-topped hill', or 'the settlement associated with *Cnossa*'.

Langley (deserted)   433235
Langeleia 1186 (British Museums MS)
OE *lang* 'long' + OE *leah* 'woodland, woodland clearing'.
'The long stretch of woodland'.

Laughton   660890
Lachestone 1086 (Domesday Book)
OE *leac* 'leek(s) + OE *tun* 'farmstead, village, small estate'.
'The place where leeks are grown, a vegetable garden'. This name has become attached to a settlement which developed on land which was once used to grow vegetables.

Launde (deserted)   797044
Landa 1155 (Derbyshire Charters)
OFr. *launde* 'open space in woodland, a forest glade'.

Present-day Launde Abbey is built on the site of an Augustinian priory founded in 1125 which was largely destroyed at the Reformation. Much of the present house was built in the seventeenth century using materials from the priory.

Leesthorpe (deserted)   792136

Luuestorp 1086 (Domesday Book)

OE ***Leof*** (male personal name) + OScand. ***thorp*** 'secondary, dependent, outlying farmstead, village'.

'The outlying farmstead associated with ***Leof***. Leesthorpe was assessed with Pickwell at Domesday.

Leicester   587046 (Clock Tower)

The earliest form of the name Leicester first appears in the written record in 787 when a bishop named in a royal charter is described as ***Legorensium episcopus*** – Bishop of ***Legorensis***, Leicester. Over the next five hundred years the name gradually changed to the form of today. Some of the early forms of the name include ***Legorensis Civitatis*** (803), ***Ligeraceaster*** (925), ***Legraceastre*** (1000), ***Ledecestra*** (1086), ***Leircestre*** (1185), ***Laycestre*** (1230) and ***Leysseter*** (1310). Not until 1610 does the name stabilise to ***Leicester***, the form in use today.

Leicester was established as a town by the Romans but there was already an important Iron Age settlement covering about 40 acres. The Romans called the town they developed ***Ratae Corieltauvororum***. The word ***Ratae*** probably derives from the Celtic word ***rath*** which means an earthen rampart or fortification. A Roman milestone found at Thurmaston dated AD 120 is engraved ***A RATIS II*** 'to *Ratis* two miles'. *Corieltauvororum* is the Latin form of the name of the Iron Age tribe in whose territory Leicester lay. Nothing remains of this Iron Age settlement but finds discovered during the excavations of the Roman bath site can be seen in Jewry Wall museum as can the Roman milestone.

The exact etymology of the 'Lei' element of the name Leicester is uncertain. Early spellings suggest that 'Lei' is a noun which derives from a previous name of the River Soar, probably the Celtic word ***Ligore*** or ***Legore,*** the name of the people who lived here. Leire, a village lying 10 miles south of Leicester on a tributary of the River Soar has the same meaning as Leicester. At Domesday Leire is listed as ***Legre***.

The second element of the name is OE ***ceaster*** a word the Anglo-Saxons used for places they recognised as Roman fortifications. **Ceaster** names are found all over the country. The most likely interpretation of the name Leicester is 'the fortified Roman town of the folk called *Legore* or *Ligore*'.

Leicester Forest

Hereswode 1086 (Domesday Book)

This forest stretched westwards from Leicester, its old name **Hereswode** means the OE **wudu** 'wood or forest' + OScand. **here** 'army, host, multitude'. This is probably a reference to the Viking army which was based in Leicester during much of the second half of the ninth century. **Herr** can also mean 'the whole people' or 'the common people'. If this were the case then this name would mean 'the wood of the common people'. The 'wood of the army' is the more likely meaning.

Leicester Frith

Frith 1322 (Calendar of Patent Rolls)

Frith is from the OE **fyrhth** 'wooded countryside' its meaning changing by the fourteenth century to 'park'. A park at this time was land which had been enclosed for hunting and which was usually enclosed by a 'pale' (bank and ditch) constructed in a special way that allowed deer to leap in but not out. Today this area is known as New Parks. The 'new park' is recorded first in 1550 as *parcus de Frithe Leicestre alias the Newe Parke of Birdesnest*.

Leire   526904

Legre 1086 (Domesday Book)

The village name has developed from the Celtic river name **Legre**, meaning unknown. The stream here is a tributary of the River Soar which flows through Leicester. **Legre** is a later name of the Soar (see Leicester). Leire is near the source of the Soar/**Legre**.

Lilinge (lost)

Lilinge 1086 (Domesday Book)

'The estate or land of **Lilla's** folk or people'.

Lilinge is one of three completely lost Domesday settlements in Leicestershire, that is, their exact sites have never been found (the other two are Netone and Plotele). Lilinge lay somewhere in the Bitteswell/Lutterworth area of Guthlaxton Hundred. This is an Anglo-Saxon name dating from the earliest years of Anglo-Saxon settlement that would have originally applied to a much wider territory than a single settlement. There is a Lilbourne about 7 miles to the south of

Lutterworth, just over the county boundary in Northamptonshire, with which there might have been a connection.

Lindley (deserted)   377948
Lindle 1209–35 (Episcopal Registers)
OE *lin* 'flax' + OE *leah* 'wood, woodland clearing, glade'.
'The woodland clearing where flax is grown'.

Lindridge (deserted)   473047
Lindrich 1316 (Leicester Corporation Deeds)
OE or OScand. *lind* 'lime tree + OE *ric* 'straight strip of raised ground'.
'A strip of land growing with lime trees'.

Little Stretton   669002
(see Great Stretton)
Little and Great Stretton were once part of the same land unit with the Great and Little being added much later to distinguish one from the other when they become separate parishes in the thirteenth century.

Littlethorpe   540968
Torp 1086 (Domesday Book)
OScand. *thorp* 'outlying secondary farmstead, village'.
Litilthorpe first appears in the records in the early fifteenth century. The primary settlement is Narborough.

Lockington   468280
Lokintone 1130 (Leicestershire Survey)
Loc (OE male personal name) + OE *ing-tun* 'farmstead, village, small estate'.
'The settlement associated with *Loc*'.

Loddington   790023
Ludintine 1086 (Domesday Book)
Ludda (OE male personal name) + OE *ing-ton* 'farmstead, village, small estate'.
'The settlement associated with *Ludda*'.

Long Clawson   725274
Clachestone 1086 (Domesday Book))

Klak (OScand. male personal name) + OE **tun** 'farmstead, village, small estate'
'The settlement associated with **Klak**'.

Long Whatton   476236
Wacton 1130 (Leicestershire Survey)
OE **wacu** 'a watch' OE **tun** 'farmstead, village, small estate'.
The meaning of this name is not clear but the most likely explanation is that **wacu** refers to 'a look-out'. The village here is near to high ground on which a look-out could have been sited.

Loughborough   534193
Lucteburne 1086 (Domesday Book)
Luhhede (OE male personal name) + OE **burh** 'fortified place'.
'The fortified place/stronghold associated with **Luhhede**'.

Lount   388195
Lunda 1347 (*Inquisitiones Post Mortem*)
OScand. **lundr** 'a small wood, a grove'.
In the Scandinavian homelands **lundr** was sometimes used for 'a sacred grove' or 'a grove offering sanctuary'. There is no evidence here for that interpretation.

Lowesby (deserted)   725078
Glowesbi 1086 (Domesday Book)
Lausi (OScand. male personal name) + OScand. **by** 'farmstead, village'.
'The settlement associated with **Lausi**'. A small new settlement has grown up here in a slightly different position.

Lubbesthorpe (deserted)   541011
Lupestorp 1086 (Domesday)
Lubb (OE male personal name) + OE **torp** or OScand. **thorp** 'outlying dependent farmstead or village'.
'The outlying settlement associated with **Lubba**'.

Lubenham   705874
Lobenho 1086 (Domesday Book)
Lubba (OE male personal name) + OE **hoh** 'spur of land'.
'The spur of land associated with **Lubba**'. There are in fact two spurs of land here.

Lutterworth   543847
Lutresurde 1086 (Domesday Book)
OE ***hluttor*** 'clear, pure, bright' + OE ***worth*** 'enclosure, farmstead, village'
This is probably ***Hultre***, an earlier name of the River Swift, 'a clear, bright stream'.

Marefield and North Marefield   752088
Merdefelde 1086 (Domesday Book)
OE ***meard*** 'marten(s)' + OE ***feld(e)*** 'open country'.
'The open country where martens abound'.

Open unencumbered ground was the original OE meaning of ***feld(e),*** by the end of the tenth century documentary evidence shows that ***feld*** had changed to mean arable, ploughed land and eventually enclosed land. This change in meaning seems to have been brought about by a steady encroachment on open land for arable farming. The Viking invasions of the second half of the ninth century and the settlement of the army along with other Viking settlers would have increased the pressure upon open land.

North Marefield is deserted.

Market Bosworth   405030
Boseworde 1086 (Domesday Book)
Bosa (OE male personal name) + OE ***worth*** 'farmstead, enclosure'.
'The enclosure associated with ***Bosa***'.
'Market' was added in the sixteenth century to distinguish this place from Husbands Bosworth.

Market Harborough   734873
Hauerberga 1177 (Pipe Rolls)
OE ***haefera*** or OScand. ***hafri*** 'oats' + OE ***beorg*** 'a hill, a mound'.
'The hill where oats are grown'.
Market Harborough did not exist until the middle of the twelfth century when it was deliberately founded as a market town. Prior to this the land that comprises present day Harborough lay in the outlying fields of Great Bowden. The hill to which the name refers is probably the hill on the A6 just north of the town.

Markfield   489100
Merchenefeld 1086 (Domesday Book)

OE **Merce** or **Mercna** "the Mercians' + OE **feld** 'tract of open country without thorns or overgrowth'.

'The open country of the Mercians'.

The most likely explanation of this name is that at the time of early Anglo-Saxon settlement this was a stretch of open land which formed a boundary between the Middle Angles and the Mercians. By the mid-eighth century the Middle Angles had been subsumed into Greater Mercia.

Feld later developed to mean land for pasture or cultivation The modern sense of the word 'field' meaning 'an enclosed or fenced-in plot of land' seems to have first come into use in the fourteenth century when the huge population decline after the various visitations of the Black Death led to large tracts of land being enclosed for pasture. The Parliamentary enclosure of arable land in the eighteenth and nineteenth centuries resulted in the strict limitation of meaning to the modern meaning of field.

Measham   335125

Messeham 1086 (Domesday Book)

This is the river name which derives from OE **meos** 'bog, marsh, moss' + OE **ham** farmstead, village'.

'The village on the River Mease'. The fields around the town are still marshy.

Medbourne   799928

Medburna 1076 (Dugdale)

Medburne 1086 (Domesday Book)

OE **maed** 'meadow' + OE **burna** 'stream'.

'The meadow stream'. A broad stream, the name of which is not known, runs through this village.

Melton Mowbray   753191

OScand. **medal** 'middle' + OE **tun** 'farmstead, village, small estate'.

The OScand. **medal** is almost certainly a replacement of the OE **middel**. The name literally means the middle farmstead or village between others. In this case there is almost certainly a more subtle meaning to the name. Melton was the central place in Framland Hundred in both in the Anglo-Saxon and Viking periods. At Domesday it was the second most important and highly valued place in Leicestershire. The church is large and has several dependent chapelries. **Middell** or **medal** is probably used here with the sense of 'central', that is, 'most important'.

Mowbray appears for the first time in the record as *Mubray* and *Moubray* in 1282. *Rogerius de Moubray* held the manor in the early twelfth century.

Misterton (deserted)   556840
Mynstretone 1086 (Domesday Book)

OE **mynster** 'minster' + OE **tun** 'farmstead, village, small estate'.

'The settlement with the minster church'.

The church here would have been an Anglo-Saxon missionary centre supporting dependent chapelries in settlements in the surrounding countryside.

Moira   318158
Moira 1831

Moira lies in the parish of Ashby-de-la-Zouch which was the property of the Earl of Moira. Fireclay and coal were found here in the mid-nineteenth century after which the town was developed.

Mountsorrel   581150
Munt Sorel 1152 (British Museum Mss)

OFr. **mont** 'a hill' + OFr. **sorel** 'sorrel'.

'The sorrel-coloured hill'.

Sorrel is a pink colour similar to the colour of the granite of this hill. The name is unlikely to refer to the sorrel plant as this would only apply for a few weeks of the year. There does not appear to have been any settlement here until a Norman castle was built in the early twelfth century.

Mowsley   647890
Muselai 1086 (Domesday Book)

OE **mus** 'mouse' + OE **leah** 'woodland clearing, wood pasture'.

'The woodland clearing infested with mice'. This does seem to be an unlikely name to have become permanently attached to a place but this type of name is common in place-names (see Foxton, Marefield).

Muston   827378
Moston 1125 (Leicestershire Survey)

OE **mus** 'mouse' + OE **tun** 'farmstead or village'.

'The settlement overrun with mice'.

Nailstone 418072

Nayllestone 1209–35 (Episcopal Registers)

Naegl (OE male personal name) + OE *tun* 'farmstead, village, small estate'.

'The settlement associated with *Naegl*'.

Naneby (deserted) 433026

Nauenebi thirteenth century (Leicestershire Charters)

Nafin (OScand. male personal name) + OScand. *by* 'farmstead, village'.

'The settlement associated with *Nafin*'.

Although this place does not appear in the records until the thirteenth century it is likely that it was already in existence by the time of Domesday but was recorded under another entry.

Nanpantan 505173

Nanpanton 1754 (map)

This name appears to refer to one building only. The meaning of this name is unclear but the most likely explanation is that it refers to the occupier of the building when the map was drawn. If this is the case it is an interesting example of the chance factors which lead to names, particularly personal names, becoming attached to places.

Narborough 542975

Norburg 1156

OE *north* 'north, northern' + OE *burh* 'a fortified place'.

'The north(ern) fortified place'.

Narborough lies to the north of Croft which was an important Anglo-Saxon royal centre in the early ninth century and probably in earlier centuries also. The absence of Narborough in Domesday Book does not mean that it did not exist it might have been included in the entry for another place but it is not clear which. The church at Narborough is a mother church with dependent chapelries an indication of considerable importance in the Anglo-Saxon period.

Near Coton 393024

(see Far Coton)

Nether Broughton 695257

Broctone 1086 (Domesday Book)

OE *broc* 'brook' + OE *tun* 'farmstead, village, small estate'.

'The farmstead or village by the brook'. Nether (lower) has been added to distinguish this village from Upper Broughton, which is now in Nottinghamshire but would once have been one single land unit with Nether Broughton.

Netone (lost)
Netone 1086 (Domesday Book)

The first element might be OE *neat* 'cattle' or it could be OE *cneo* a 'knee or 'bend'. This could be either the *tun* '?settlement on the bend of a river or road' or 'a cattle farm'.

Netone is one of three settlements recorded in the Domesday Book for Leicestershire which are lost completely, that is it is not known where they lay, the others are Lilinge and Plotele. The names have an unfamiliar ring to them as they were frozen at an early point in their development. Netone was sited in Gartree Hundred in the Cold Newton/Burrough-on-the-Hill area.

Nevill Holt   817938
Holt 1150 (The Red Book of the Exchequer)
OE *holt* 'a wood, a thicket'.

'The wood'. The name *holt* was often attached to a single-species wood. In 1498 this manor was held by Thomas *Neville*.

New Parks (Leicester)
Sixteenth century

This was once part of Leicester Forest which had been cleared by the sixteenth century when it first appears in the records as a separate unit. It is now almost entirely covered with houses.

Newbold (near Owston)   401190
Neubotel/Neubold 1130 Leicestershire Survey
OE *niew* 'new' + OE *botl* 'building, dwelling'.
'The new building'.

Newbold Folville (deserted)   706120
Neubold 1086 (Domesday Book)
OE *niwe* 'new' + OE *bold/botl* 'building'.

'The new building'. This might have been a new important house. If the name referred to a lesser building the element cottage(s) would have been more appropriate. *Willelmus de Folevill* held this manor in 1236, this is the same family who held the manor of Ashby Folville.

Newbold Saucy (deserted)   765090
Newboldia Henry II 1154-1189 (Danelaw Charters)
OE *niwe* 'new' + OE *bold/botl* 'building'.
'The new building'. Held by Robert *de la Sauce*. This is probably Saussay in Manche, France.

Newbold Verdon   446038
Niwebold 1086 (Domesday Book)
OE *niwe* 'new' + OE *bold/botl* 'build, building, dwelling'.
'The new building'. Held by *Nicholaus de Verdon*, from Verdun in France, held the manor in 1226.

Newbold (Worthington)   402194
Neubold 1212 (Feet of Fines)
OE *niwe* 'new' + *bold/botl* 'building'.
'The new building'. This small medieval settlement was deserted in the sixteenth century and a new settlement has grown up in modern times.

Newton Burgoland   370090
Neutone 1086 (Domesday Book)
OE *niwe* 'new' + OE *tun* 'farmstead, village, small estate'.
'The new settlement'. This manor was held by Roger *de Burgylum* in 1225.

Newton Harcourt   637970
Niwetone 1086 (Domesday Book)
OE *niwe* 'new' + OE *tun* 'farmstead, village, small estate'.
'The new settlement'. This manor was held by Richard de Harcourt in 1240.

Newtown Linford   517104
Neuton 1325 (*Inquisitiones Post Mortem*)
OE *niwe* 'new' + OE *tun* 'farmstead, village, small estate'.
Lindenforth 1327 (SR) + OE *linden* 'lime-tree' + OE *ford* 'ford'
'The new settlement at the ford where the lime-trees grow'. This place was settled from Groby.

Newtown Unthank   489044
Neuton 1282 (Index of *Placita de Banco*)
OE *niwe* 'new' + OE *tun* 'farmstead, village, small estate'.

'The new settlement'.

Unthank is a family name, a Robert *Unthanke* appears in the records of Kirby Muxloe Castle in 1481.

Normanton (deserted) 811406

Normanton 1130 (Leicestershire Survey)

OE *northman* 'Norwegian' + OE *tun* 'farmstead or village, small estate'.

The settlement of the Norwegians. This name would have arisen to distinguish the people of this settlement from other Scandinavian groups, probably Danes, in the area. This settlement lay one mile north of Bottesford.

Normanton-le-Heath 377128

Normenton 1247 (Fine)

OE *Northman* 'Norwegian' + OE *tun* 'farmstead, village, small estate'

'The settlement of the Norwegian(s)'.

This name reveals the existence of an identifiable distinctive ethnic group which suggests that Norwegian Vikings were an exception in this area. The heath here stretched to the edge of Charnwood Forest.

Normanton Turville (deserted) 489995

Normanton 1191(Rotuli Hugh de Welles)

OE *Northman* 'Norwegian' + OE *tun* 'farmstead, village, small estate'.

The settlement of the Norwegians'.

This name indicates that there was a distinctive group of Norwegian Vikings settled here. The *de Turville* family held the manor here from the early thirteenth century until 1547.

North Kilworth 616834

Chivelesworde 1086 (Domesday)

Cyfel (OE male personal name) + OE *worth* 'enclosure, farmstead, village'.

'The settlement associated with the people of *Cyfel*'.

North Marefield (deserted)

(see Marefield)

Norton-juxta-Twycross 324070

Nortone 1086 (Domesday Book)

OE *north* 'north' + OE *tun* 'farmstead, village, small estate'.

'The north settlement'. This type of place-name usually indicates that the settlement is north of a more important place often one to which it is attached. Juxta begins to be used after 1327 to distinguish this place from other places which have the same name.

Noseley (deserted) 733987

Nothwulf (OE male personal name) + OE *leah* 'woodland, woodland clearing, wood pasture'.
'The woodland pasture associated with *Nothwulf*'.

Oadby 625005

Oldebi 1086 (Domesday Book)
Audi (OScand. personal name) + OScand. *by* 'farmstead, village'.
'The settlement associated with *Audi*'. This is a Viking name which must have replaced an earlier name as an Anglo-Saxon cemetery has been found at Brocks Hill.

Oakthorpe 320129

Achetorp 1086 (Domesday Book)
Aki (OScand. male personal name) + OScand. *thorp* 'outlying or dependent settlement'.
'The outlying settlement associated with *Aki*'.

Odstone 393079

Odestone 1086 (Domesday)
Oddr (OScand. male personal name) + *tun* 'farmstead, village, small estate'.
'The settlement associated with *Oddr*'. This Viking personal name compounded with OE *tun* indicates that there was a prior Anglo-Saxon settlement here.

Old Dalby 673238

Dalbi 1086 (Domesday Book)
OScand. *dalr* 'dale, valley' + OScand. *by* 'farmstead, village'.
'The settlement in the valley'. Old is a form of OE *wald* 'woodland, a tract of high forest land'. With the clearing of large forest tracks, some of which were on high ground, wold came to describe high open land which had once sustained trees.

Orton-on-the-Hill 306038

Wortone 1086 (Domesday Book)
OE *uferra* 'higher, upper' + OE *tun* 'farmstead, village, small estate'.

'The upper settlement'. Orton is sited on the crest of a high spur of land. 'On the Hill' occurs for the first time at the end of the sixteenth century.

Osbaston   425045

Sbernestun 1086 (Domesday Book)

Asbjorn (OScand. male personal name) + OE *tun* 'farmstead, village, small estate'.

'The settlement associated with *Asbjorn*'.

This name is a Viking personal name combined with the Anglo-Saxon *tun*, an indication that this is a renaming by an incomer, or usurper, of a previously existing settlement.

Osgathorpe   430196

Osgodtorp 1086 (Domesday Book)

Asgot (OScand. personal name) + OScand. *thorp* 'outlying, dependent farmstead, village'.

The settlement associated with *Asgot*'.

Othorpe (deserted)   771955

Actorp 1086 (Domesday Book)

Aki (OScand. male personal name) + OScand. *thorp* 'outlying, dependent farmstead or village'.

'The settlement associated with *Aki*'.

Owston   776078

Osulvestone 1086 (Domesday Book)

Oswulf (OE male personal name) + OE *tun* 'farmstead, village, small estate'.

'The settlement associated with *Oswulf*'.

Packington   362147

Packynton 1050 (*Diplomatarium Anglicum*)

Pachingtone 1086 (Domesday Book)

Pacca (OE male personal name) + OE tun 'farmstead, village, small estate'.

'The settlement associated with *Pacca*'.

Peatling Magna   595927

Petlinge 1086 (Domesday Book)

Peotla (OE male personal name)

'The settlement of **Peotla's** people'. This is a folk or kin name which dates from an early period of Anglo-Saxon settlement in the county (sixth/seventh centuries).

Magna appears for the first time in 1224. The two Peatlings would once have been one land unit.

Peatling Parva   589897

(see Peatling Magna)

alia Petlinge 1086 (Domesday Book)

'The other Petlinge'.

Parva is first recorded in 1226.

Peckleton   472011

Pechintone 1086 (Domesday Book)

Peohtla (OE male personal name) + OE *tun* 'farmstead, village, small estate'.

'The settlement associated with **Peohtla**'.

Peggs Green (Coleorton)   414175

This place is not recorded until 1795 in Nichols. There is no earlier reference to this place the meaning of which is obscure but most probably refers to a particular person. See Nanpantan for a similar example.

Pickwell   784113

Picheuuelle 1086 (Domesday Book)

OE *pic* 'peak' + OE *wella* 'spring or stream'.

The spring or stream near the peak(s).

There are pointed hills to the south-east of Pickwell and the village is sited near the head of a stream.

Pinwall   308999

Pynvell 1346 (Nichols)

Pinna (OE male personal name) + OE *wella* 'spring or stream'.

'The stream or spring associated with **Pinna**'. Because this name appears rather late in the record it is possible that **Pinna** may not be a personal name.

Plotele (lost)

Plotele 1086 (Domesday Book)

OE *plot* 'plot of land' + OE *leah* 'woodland glade'.

'A plot of land in a woodland glade'.

Plotele was one of three completely lost Domesday settlements in Leicestershire, that is, their exact sites have never been found. The other two are Netone and Lilinge. It is known that Plotele was in Guthlaxton Hundred.

Plungar 768339

Plungar 1130 (Leicestershire Survey)

OE *plume* 'a plum-tree' + OE *gara* 'a gore, a triangular plot of land' or OScand. *garth* 'an enclosure'.

'An enclosed corner of land growing with plum trees'.

Potters Marston 498964

Mersitone 1086 (Domesday Book)

OE *mersc* 'marsh' + OE *tun* 'farmstead, village, small estate'.

'The settlement near the marsh'. By the thirteenth century *Poteresmerston* appears in the records an indication that that this settlement was known as a centre for pottery making.

Poultney (deserted) 586848

Pontenei 1086 (Domesday Book)

Pulta (OE male personal name) + OE *eg* 'island, dry land in well-watered country'.

'Pulta's island'.

Prestgrave (deserted) 801937

Abbegrave 1086 (Domesday Book)

Prestgrave 1130 Leicestershire Survey

OE *abba* 'abbot' OE *preost* 'priest' + OE *graf* 'coppiced/managed wood'.

'Abbot's wood, later becoming priest's wood'. Prestgrave lay next to Holt, another tract of managed woodland.

Prestwold (deserted) 580216

Prestewalde 1086 (Domesday Book)

OE *preost* 'priest' + OE *walde* 'woodland, tract of woodland, high forest land'.

'The priest's wood or woodland'. The product of this *wald* would have supported a priest (or priests).

Primethorpe (deserted) 523930

Torp 1086 (Domesday Book)

Prim (OE male personal name) + OScand. *thorp* 'dependent outlying farmstead, village' .

'The settlement associated with **Prim**'. This place has been subsumed into Broughton Astley.

Quenby (deserted)   702065

Queneberie 1086 (Domesday Book)

OE *cwen* 'queen' + either OScand. *by* 'farmstead, village' or OE *burh* 'fortified place'.

'The queen's fortified place' or the 'queen's settlement'. The early spellings of this name do not make it clear as to whether the second element is *by* or *burh*. The circumstances which would have led the element *cwen* to be coined would not have applied in the Viking period, making *burh* the more likely element. Over time the original meaning of the name was lost with *by* replacing *burh* because of association with the other *bys* in the area.

Queniborough   648122

Cuinburg 1086 (Domesday Book)

OE *cwen* 'queen' + OE *burh* 'fortified manor/settlement'.

'The queen's fortified settlement'. This is the same name as Quenby, these two places are nor far apart with both names referring to a now unknown royal landowner.

Quorn(don)   565165

Querendon 1154–89 (Henry II MS cited by Dugdale)

OE *cwearn* 'quern' + OE *dun* 'a hill, an expanse of open hill country'.

'The hill where querns are got'.

Ragdale   622199

Ragendele 1086 (Domesday Book)

OScand. *rake* 'narrow path up a ravine' + OE/OScand. *dael* 'valley, pit, hollow'.

The topography here confirms the meaning of this name with a valley rising up from the River Wreake.

Ratby   513060

Rotebie 1086 (Domesday Book)

Rota (OE male personal name) + OSCand. *by* 'village, farmstead'.

'The settlement associated with **Rota**'.

This is an interesting name in that it comprises an early Anglo-Saxon man's name with the OScand. word **by**, a strong indication that there was an already existing Anglo-Saxon settlement here before the coming of the Vikings. A **Rota** also gave his name to Rutland but there does not appear to be any connection between the two.

There is another possible, but so far unprovable, interpretation of this name. The first element of this name could be the Celtic word **rath** 'a defensive entrenchment' and the late Iron Age earthwork here, known as Ratby Burrows, certainly fits that description. It is also possible that the name **rath** was transferred to the name of Roman Leicester, **Ratae Corieltauvorum**, which was an important Iron Age site before the Romans establish their *Civitas* centre there.

Ratcliffe Culey 326995

Redeclive 1086 (Domesday Book)

OE **raed** 'red' + OE **clif** 'bank or cliff'.

'The red cliff or bank'. This must refer to the red marl that can be seen on the river bank. The cliff may have referred to a steep bank along the river as there is no other obvious cliff here. In 1228 this manor was held by *Johannes de Cuylly*.

Ratcliffe-on-the-Wreake 630145

Redeclive 1086 (Domesday Book)

OE **raed** 'red' + OE **clif** 'cliff'.

'The red cliffe'. The geology here is red marl. '*Super le Wrethek*' appears in the record for the first time in 1259.

Ravenstone 405137

Ravenestorp 1086 (Domesday Book)

Hraefn (OScand. male personal name) + OE **tun** 'farmstead, village, small estate'.

'The settlement associated with **Hraefn**'.

Rearsby 650144

Redresbi 1086 (Domesday Book)

Reitharr (OScand. male personal name) + OScand. **by** 'farmstead, village'.

'The settlement associated with **Reitharr**'. It is possible that this is the same **Reitharr** who held nearby Rotherby.

Redmile 788355

Redmelde 1086 (Domesday Book)

OE *raed* 'red' + OE **mylde** 'soil, earth'.
'The place with red earth'.

Ringolthorpe (deserted) 776235
Ricoltorp 1086 (Domesday Book)
Ringulfr (OScand. male personal name) + OScand. **thorp** 'outlying dependent farmstead or village'.
'The settlement associated with **Ringulfr**'.
'Goldsmiths Grange', named after John Goldsmith who lived there in the first half of the fifteenth century, now stands on the site of this deserted village.

Rolleston 732004
Rovestone 1086 (Domesday Book)
Hrolfr (OScand. male personal name) + OE **tun** 'farmstead, village, small estate'.
'The settlement associated with **Hrolf**'. The Viking personal name and Anglo-Saxon **tun** indicate that this is a renaming, probably by a Viking usurper, of an already existing Anglo-Saxon settlement.

Rotherby 676167
Redebi 1086 (Domesday Book)
Reitharr (OScand. male personal name) + OScand. **by** 'farmstead, village'.
'The settlement associated with **Reitharr**', who may also have given his name to nearby Rearsby.

Rothley 585126
Rodolei 1086 (Domesday Book)
OE **roth** 'a clearing' + OE **leah** 'woodland, woodland clearing'.
'A woodland clearing'.

Saddington 659919
Sadintone 1086 (Domesday Book)
Saegeat (OE male personal name) + OE **tun** 'farmstead, village, small estate'.
'The settlement associated with **Seageat**'.

Saltby 851263
Saltebi 1086 (Domesday Book)
OE **salt** 'salt' + OScand. **by** 'farmstead, village'.

'The farmstead or village at the salt spring'.

There is a chalybeate spring just to the south-west of the village. It is possible, but unlikely given the existence of a major salt spring nearby, that **Salt** could be an OScand. male personal name.

Sapcote 490935

Scepecote 1086 (Domesday Book)

OE **sceap** 'sheep' + OE **cot** 'cottage, shelter'.

'The shelter for sheep'.

Saxby 820200

Saxebi 1086 (Domesday Book)

This could be either the OScand. personal name **Saxi** + OScand. **by** farmstead or village, or the OE folk name **Seaxe**.

Either 'the settlement associated with **Saxi**' or 'the settlement of the Saxons'.

If the meaning is the latter then there must have been something rather significant about the place. It is possible that there was a strong Anglo-Saxon presence here at the time of the Viking settlement of the area. There are several other settlements in this area that have an OE first element with the OScand. suffix **by.**

Saxelby 700210

Saxelbie 1086 (Domesday Book)

Saksulfr (OScand. male personal name) + OScand. **by** 'farmstaed, village'.

'The settlement associated with **Saksulfr**'.

Scalford 763242

Scaldeford 1086 Domesday Book)

OE **sceald** 'shallow' + OE **ford** 'ford'.

'The shallow ford'.

The OE **sce** (pronounced 'sh') has been replaced by the OScand. **sk** with a hard 'k', an indication of a Viking presence here.

Scraptoft 648058

Scrapentot 1086 (Domesday Book)

Skrapi (OScand. male personal name) + OScand. **toft** 'A small plot of land'.

'The small plot of land associated with **Skrapi**'.

Seagrave   620175
Satgrave 1086 (Domesday Book)
OE **seath** 'pit, pool' + OE **graf** 'grove'.
'The pit or pool at the grove'.

Sewstern   889218
Sewesten 1086 (Domesday Book)
Saewig (OE personal name) + OE **thyrne** 'thorn'.
'Thorn or thorny land associated with **Saewig**'.

Shackerstone   374068
Sacrestone 1086 (Domesday Book)
OE **sceacere** 'robber' + OE **tun** 'farmstead, village, small estate'.
'The settlement where there are robbers'.
There is no evidence as to how this name arose but banditry would not have been rare in Anglo-Saxon and Viking times.

Shangton   717962
Santone 1086 (Domesday Book)
OE **scanca** 'shank, leg, narrow spur of a hill' + OE **tun** 'farmstead, village, small estate'.
'The settlement near a narrow ridge of land'.

Sharnford   482919
Scearneforde 1086 (Domesday Book)
OE **scearn** 'dung, muck' + **ford** 'ford'.
'The mucky ford'. **Scearn** is often combined with brooks, streams, fords. In dry weather a shallow ford would quickly become a filthy mess.

Shawell   546804
Sawelle 1086 (Domesday Book)
OE **sceath** 'boundary' + OE **wella** 'spring or stream'.
'The stream which crosses the boundary'.
A stream crosses the Roman Watling Street here, an ancient boundary even before the Romans came.

Shearsby   624909
Seuesbi 1086 (Domesday Book)
Skeifr (OScand. male personal name) + OScand. **by** 'farmstead, village'.

'The settlement associated with **Skeifr**'.

Sheepy Magna   327015
Scepa 1086 (Domesday Book)
OE s*ceap* 'sheep' + OE *eg* 'island, piece of dry land in fenny country'.
'The dry island of land, used for grazing sheep, in a fen'. Magna was
added to distinguish this place after 1276.

Shelthorpe   545185
Serlesthorp 1284 (Assize records)
Serlo (OScand. male personal name) + OScand. *thorp* 'outlying,
dependent farmstead, hamlet'.
'The settlement associated with **Serlo**'.

Shenton   386001
Scenctune 1102 (Anglo-Saxon will)
Scenctun 1086 (Domesday Book)
OE *scence* 'a draught, a drink' + OE *tun* 'farmstead, village, small
estate'.
'The settlement on the River Sence'. Shenton stands on a tributary of
the River Sence known locally as Sence Brook.

Shepshed   478194
Scepeshefde 1086 (Domesday Book)
OE *sceap* 'sheep' + OE *heafod* 'hill, head'.
'The hill where sheep graze'.

Shoby   683203
Seoldesberie 1086 (Domesday Book)
Sigvaldi (OScand. male personal name) + OScand. *by* 'farmstead,
village'.
The settlement associated with **Sigvaldi**'.

Sibson   355009
Sibetesdone 1086 (Domesday Book)
Sigebed (OE male personal name) + OE *dun* 'a hill, an expanse of
open hill country'.
'The hill associated with **Sigebed**'.

Sileby   608151
Siglesbie 1086 (Domesday Book)

Sigulfr (OScand. male personal name) + OScand. ***by*** 'farmstead, village'.

'The settlement associated with ***Sigulfr***'.

Six Hills   644207

1795 (Nichols)

This place was recorded as Segeswold in 1156. ***Seges*** is the OE male personal name ***Sex*** or ***Seg*** + OE w***ald*** 'large tract of woodland, high forest land' which after clearing described 'high open land'. This is a good description of the landscape here.

Segeswold changed over time to become Six Hills, which is at the crossroads of the Roman Fosse Way and a prehistoric 'Salt Way'. The boundaries of six parishes also meet at this crossroads.

Skeffington   743028

Sciftitone 1086 (Domesday Book)

Sceaft (OE personal name) + OE ***inga-tun*** 'farmstead, village, small estate'

'The estate of ***Sceaft's*** people'. This type of ***inga-tun*** name comes from the period of early Anglo-Saxon settlement when the people would have settled in kin groups with their leader, in this case ***Sceaft***.

Sketchley   425923

Skettesley 1558

The meaning of this name is unclear as the first time it is recorded is too late for a secure interpretation. It could be either OE ***Sket*** or OScand. ***Skiotr*** (male personal names) or OScand. ***skeith*** 'boundary road, boundary' + OE ***leah*** 'wood, woodland clearing, glade'. The interpretation of the first element as ***skeith*** fits rather nicely with the Roman road, Watling Street, forming the south-west boundary of this settlement and the county.

Slawston   778944

Slagestone 1086 (Domesday Book)

Slagr (OScand. male personal name) + OE ***tun*** 'farmstead, village, small estate'.

'The settlement associated with ***Slagr***'.

Smeeton Westerby   678928

Smiteton 1086 (Domesday Book)

Westerbi 1206 (Curia Regis Rolls)

These two separate settlements are now treated as one village.

OE *smethe* 'smith' + OE *tun* 'farmstead, village, small estate'.

OScand. *vestra* 'west, westerly' + OScand. *by* 'farmstead, village'.

'The settlement of the smiths' and the 'western settlement' (to the west of Smeeton).

The first recorded instance of the two settlements fused as one is in 1316.

Smockington   453899

Snochantone 1086 (Domesday Book)

The first element could be the OE male personal name *Snocca* or it could be OE *snoc* 'point, projection' + OE *ing-ton* 'farmstead, village, small estate'.

'The settlement near the promontory' or 'the settlement associated with *Snocca*'.

Snareston   343094

Snarchestone 1086 (Domesday Book)

OE *Snarc* (male personal name) + OE *tun* 'farmstead, village, small estate'.

'The settlement associated with *Snarc*'.

Snibston   414147

Snipeston 1200 (*Placitorum Abbrevatio* 1811)

Snipr (OScand. male personal name) + OE *tun* 'farmstead, village, small estate'.

'The settlement associated with *Snipr*'. The Viking name almost certainly replaced an earlier Anglo-Saxon name.

Somerby   779105

Sumerlidebie 1086 (Domesday Book)

Sumarlidi (OScand. male personal name) + OScand. *by* 'village, farmstaead'.

'The place associated with *Sumarlidi*'.

South Croxton   690103

Crochestone 1086 (Domesday Book)

Krokr (OScand. male personal name) + OE *tun* 'farmstead, village, small estate'.

'The settlement associated with *Krokr*'.

This name combines a Viking personal name with the OE *tun*, an indication that an older Anglo-Saxon settlement has been taken over by a Viking incomer. **Suth** 'south' first appears in the records in 1201, it would have arisen to distinguish this place from Croxton Kerrial.

South Kilworth   604819
Chivelsworde 1086 (Domesday Book)
(See North Kilworth)

Sproxton   858244
Sprotone 1086 (Domesday Book)
Sprok (OScand. male personal name) + OE **tun** 'farmstead or village'
'The place associated with **Sprok**'. There was almost certainly an Anglo-Saxon land-holder here who was usurped by **Sprok**.

Stanton-under-Bardon   467105
Stantone 1086 (Domesday Book)
OE **stan** 'stone' + OE **tun** 'farmstead, village, small estate'.
'The stone-built settlement sited at the foot of Bardon Hill'. *Undirberdon* is first recorded in 1327.

Stapleford (deserted)   813183
Stapeford 1086 (Domesday Book)
OE **stapol** 'a post' + OE **ford** 'ford'.
'The ford marked by a post'. The element **stapol** is sometimes used to indicate a boundary marker. Stapleford is on the boundary between Leicestershire and Rutland but the ford is half a mile from the boundary.

Stapleton   435986
Stapeltone 1086 (Domesday Book)
OE **stapol** 'a post' + OE **tun** 'farmstead, village, small estate'.
In place-names **stapol** usually indicates a land boundary of some kind.

Starmore (deserted)   588804
Stormeorde 1086 (Domesday Book)
Storm (OE male personal name) + OE **worth** 'enclosure, village'.
'The enclosed settlement associated with **Storm**'. Over time **worth** has mutated into 'more' when the original meaning was lost to memory.

The settlement stood on Hovel Hill opposite the site of **Westerhyll** 1578 (Braye Mss).

'The western hill', now known as Gravel Hill. Both of these villages had been abandoned by the sixteenth century.

Stathern   771310

Stachedirne 1086 (Domesday Book)

OE *staca* 'stake' + OE **thyrne** 'thorn'.

'The stake-thorn'.

It is unclear what this compound name meant exactly. It was probably used to indicate a boundary post as the OE *staca* is found in other place-names sited on boundaries.

Staunton Harold (deserted)   379209

Stantone 1086 (Domesday Book)

OE *stan* 'stone' + OE *tun* 'farmstead, village, small estate'

'The settlement built of stone'. Limestone would have been easily obtained from nearby Breedon Hill. *Harald de Stantona* held this manor in the early twelfth century.

Stockerston   838977

Stoctone 1086 (Domesday Book)

All early spellings of this place, with the exception of Domesday, are OE *stocfaston* 'a stockade, stronghold' + OE *tun* 'farmstead, village, small estate'.

'The place where the stronghold of heavy timbers is built'.

Stoke Golding   400974

Stochis 1173 (Dugdale)

OE *stocu* 'outlying dairy farm(s)'.

Petrus de Goldinton held this manor in 1200.

Stonesby   823246

Stovenebi 1086 (Domesday Book)

OE/OScand. *stofn* 'tree-stump' + OScand. *by* 'farmstead, village'.

'The settlement at or near the tree-stump(s)'.

Stoneygate (Leicester)   600033

OE *stan* 'stone' and OScand. *gata* 'road'.

Named from 'Stoneygate House' which is marked on an 1806 map of Leicester. The house stood near where the Evington Footpath leaves London Road opposite Victoria Park gates.

Stoney Stanton   490948

Stantone 1086 (Domesday Book)
OE *stan* 'stone' + OE *tun* 'farmstead, village, small estate'.
'The settlement on stony ground'.
'Stoney', which had appeared in the records by the fourteenth century, was added to distinguish this village from Stanton under Bardon and Staunton Harold.

Stonton Wyville   737951

Stantone 1086 (Domesday Book)
OE *stan* 'stone' + OE *tun* 'farmstead, village, small estate'.
This could be a 'settlement built on stoney ground' or 'the settlement built of stone'. The former is the most likely. The manor was held by Robert *de Wivill* in 1209. The name derives from Gouville in Nortmandy.

Stoughton   641011

Stoctone 1086 (Domesday)
OE *stoc* 'tree trunk, tree stump, log (epecially if left standing)' + OE *tun* 'farmstead, village, small estate'.
'The settlement where the tree stumps are'.

Stretton-en-le-field   305119

Streiton 1086 (Domesday Book)
OE *straet* 'Roman road' + OE *tun* 'farmstead, village, small estate'.
'The settlement at or near a Roman road'. The Anglo-Saxons used the element **straet** when they recognised a road as Roman in origin. There is no kown Roman road here although the village is on the line of an old routeway known as the Salt Way. The line of this road comes across the county from Grantham, through Croxton Kerrial across to Six Hills where it crosses the Roman Fosse Way, then on to Barrow upon Soar. The line is then lost but can be picked up coming across to Stretton en le Field.
'In the Field' appears for the first time in the records in the early fifteenth century. It means 'in the open country' not 'in the fields'.

Stretton Magna   657005

(see Great Stretton)

Sutton Cheyney   418005

Sutone 1086 (Domesday Book)

OE *suth* 'south' + OE *tun* 'farmstead, village, small estate'.

This settlement lies to the south of the important royal Anglo-Saxon centre of Market Bosworth. John *Chaynel* held the manor in the late thirteenth century.

Sutton-in-the-Elms   519939

Sutone 1086 (Domesday Book)

OE *suth* 'south' + OE *tun* 'farmstead, village, small estate'.

This settlement lies to the south of the important royal Anglo-Saxon centre of Croft. 'In the Elms' was added much later to distinguish this place from the other Suttons in the region.

Swannington   415160

Swaneton 1199 (Feet of Fines)

Swan (OE male personal name) + OE *ing-tun* 'farmstead, village, small estate'.

'The settlement associated with *Swan*'.

Swepstone   368105

Scopestone 1086 (Domesday Book)

Sweppi (OE male personal name) + OE *tun* 'farmstead, village, small estate'.

'The settlement associated with *Sweppi'*.

Swinford   569794

Suineford 1086 (Domesday Book)

OE *swin* 'swine/pig' + OE *ford* 'ford'.

'The swine ford'.

Swithland   549131

Svithelund 1278 (*Rotuli Rcardi*, 1925)

OScand. *svithinn* 'land cleared by burning' + OScand. *lundr* 'grove'.

'The grove cleared by burning'. In the Viking homelands the word *lund* was often used for a sacred grove.

Sysonby (deserted)   739190

Systenebie 1086 (Domesday Book)

S*igsteinn* (OScand. male personal name) + OScand. *by* 'village or farmstead'.

'The place associated with **Sigsteinn**'.

Syston   632113
Sitestone 1086 (Domesday Book)
Sigestein (OE male personal name) + OE **tun** 'farmstead, village, small estate'.
'Settlement associated with **Sigestein**'.

Theddingworth   668857
Tedingesworde 1086 (Domesday Book)
Theoda (OE male personal name) + OE **ing** + OE **worth** 'enclosure, farmstead'.
'The enclosed settlement of the followers of **Theoda**'.
This name dates from an early period of Anglo-Saxon settlement.

Thornton   468076
Torenton 1086 (Domesday Book)
OE **thorn** 'thorn' + OE **tun** 'farmstead, village, small estate'.
'The settlement where thorn-trees grow'.

Thorpe Acre   517200
Torp 1086 (Domesday)
Thorp Haueker 1343 (Papal Registers)
OE **thorp** 'outlying, dependent farmstead, hamlet' + OE **hafocere** 'hawker'.
'The outlying settlement associated with a hawker'.

Thorpe Arnold   771202
Torp 1086 (Domesday Book)
Thorp Ernad 1239 (Episcopal Registers)
OScand. **thorp** 'outlying, dependent, farmstead, village' dependent on Melton Mowbray with which it was assessed at the time of Domesday.
Ernaldi appears for the first time in the records in the early thirteenth century. *Earnald de Bosco* held the manor in 1156 followed by at least five successive generations of the family.

Thorpe Langton   742925
Torp 1086 (Domesday Book)
OScand. **thorp** 'outlying, dependent farmstead, village' of Langton.
(See Church Langton)

Thorpe Satchville   733118

Thorp 1130 (Leicestershire Survey)

OScand. ***thorp*** 'outlying, dependent farmstead, village'

This place was held by *Radulfus de Secheville* in 1210, the name derives from Secqueville in Normandy.

Thringstone   425176

Trangesbi 1086 (Domesday Book)

Thraeingr (OScand. male personal name) + OE ***tun*** 'farmstead, village, small estate'. The ***bi*** of the Domesday spelling is the OScand. ***by*** 'farmstead, village' settlement' which has, over time, changed to ***tun***. It is likely that ***by*** replaced ***tun*** when the Viking ***Thraeingr*** usurped a previous Anglo-Saxon holder with the older ***tun*** being later reinstated. A change of suffix in place-names is not uncommon.

Thrussington   649159

Turstanstone 1086 (Domesday Book)

Thorstein (OScand. male personal name) + OE ***tun*** 'farmstead, village, small estate'.

'The settlement associated with ***Thorstein***'. This hybrid name combines a Viking personal name with OE ***tun***, an indication that this is probably a renaming by Viking incomers/usurpers of an already existing Anglo-Saxon settlement.

Thurcaston   568110

Turchitelestone 1086 (Domesday Book)

Thorketill (OScand. male personal name) + OE ***tun*** 'farmstead, village, small estate'.

'The settlement associated with ***Thorketill***'. This name combines a Viking personal name with the OE suffix ***tun*** an indication that this settlement was in existence before the coming of the Vikings.

Thurlaston   504991

Turlauestona 1196 (Chancery Records)

Thorleif (OScand. male personal name) + OE ***tun*** 'settlement'.

'The settlement associated with ***Thorlief***'.

The Viking personal name combined with the OE ***tun*** indicates that this is probably a renaming of an existing Anglo-Saxon settlement by a Viking incomer/usurper.

Thurmaston   610090

Turmodestone 1086 (Domesday Book)

Thormothr (OScand. male personal name) + OE *tun* 'farmstead, village, small estate'.

'The settlement associated with *Thormothr*'.

The Viking personal name with the OE *tun* is an indication that this is a renaming of an already existing Anglo-Saxon settlement by a Viking incomer/usurper.

Thurnby   647039

Turnebi 1156

Thyrne (OScand. male personal name) + OScand. *by* 'farmstead, village'.

'The settlement associated with *Thyrnir*'.

Tilton-on-the-Hill   743056

Tillintone 1086 (Domesday Book)

Tila (OE male personal name) + OE *tun* 'farmstead, village, small estate'. 'On the Hill' is a modern addition.

Tongue   418232

Tunge 1086 (Domesday Book)

OE *tong* 'tongue, tongue of land'

Tooley (deserted)   475995

Tolawe 1133–89 (British Museum Ms., Henry II)

OE *tot-hlaw* 'look-out hill'. The OE word *hlaw* is often used for an artificial hill such as a burial mound.

Tugby   763010

Tochebi 1086 (Domesday Book)

Toki (OScand. male personal name) + OScand. *by* 'farmstead or village'.

'The settlement associated with *Toki*. At Domesday the holder of this manor is listed as *Toki* thus giving a starting date for the name of the settlement which would previously have been known by another name. It is rare to obtain a definite date for the coining of a name.

Tur Langton   714946

Terlintone 1086 (Domesday Book)

Tyrthtel (OE male personal name) + OE **tun** 'farmstead, village, small estate'.

'The settlement associated with **Tyrhtel**'.

Over time the original meaning of this name would have been forgotten with the name developed into its present form, eventually becoming Langton by association with its neighbouring villages.

Twycross   338049

Tvicros 1086 (Domesday Book)

OE **twi** 'double, two' + OE **cros** 'cross'.

OE **cros** in place-names usually refers to a carved stone cross. The name probably refers to two crosses, neither of which has survived.

Twyford   729102

Tuiuorde 1086 (Domesday Book)

OE **twi** 'double, two' + OE **ford** 'ford'.

This probably refers to two fords which lay close together across two arms of a river.

Ullesthorpe   507877

Ulestorp 1086 (Domesday Book)

Ulfr (OScand. male personal name) + OScand. **thorp** 'outlying, dependent farmstead, village'.

'The settlement associated with **Ulfr**'. This is a wholly Viking name with Ullesthorpe being settled from Claybrooke.

Ulverscroft   486117

Ulfescroft 1277 (*Rotuli Ricardi*, 1925)

Ulfr (OScand. male personal name) + OE **croft** 'small enclosed field, piece of land'.

'The land held by **Ulfr**'. An Augustinian priory was founded here in 1135 by Robert *le Bossu*, Earl of Leicester.

Upton   364997

Upton 1130 (Leicestershire Survey)

OE **upp** 'high, higher' + OE **tun** 'farmstead, village, small estate'.

'The high or higher settlement'.

Walcote   568838

Walecote 1086 (Domesday Book)

OE **walh** 'Briton, serf, slave' + **cot** 'cottage, shelter'.

'The cottages of the Welsh, Britons or serfs'. **Walh** is the name the Anglo-Saxons used for the indigenous Romano-British people, it also carries the meaning of serf or slave which was the condition to which many Romano-British people were reduced. There would have been some inter-mingling of the two groups but place-name evidence reveals these specific settlements for some of the indigenous people. These settlements might have been groups which had held out against the incomers and later became tolerated or they could have been groups who had been 'settled' by the Anglo-Saxons for some servile function.

Waltham-on-the-Wolds   803250
Waltham 1086 (Domesday Book)

OE **wald** 'large tract of woodland, or high forest land' + OE **ham** 'village, estate, homestead'. This is OE **wald-ham** as in 'a forest estate'. This name belongs to the period of early Anglo-Saxon settlement c.600 or earlier. These early Anglo-Saxon places were often near Roman roads and Waltham lies exactly on the minor Roman road known as King Street and one mile from the more major Roman road across the wolds. The name might also carry the meaning of a royal hunting estate. **Wald** can also mean a stretch of high open country or moorland.

Walton   597872
Waltone 1086 (Domesday Book)

OE **walh** 'Briton, serf/slave, Welshman' + OE **tun** 'farmstead, village, small estate'.

'The settlement of the Britons or the serfs' (See Walcote.)

Walton-on-the-Wolds   593198
Waletone 1086 (Domesday Book)

(See Walcote).

'*Super waldas*' appears in the record in 1354.

Wanlip   600109
Anlepe 1086 (Domesday Book)

OE **anliepe** 'single, solitary, lonely'.

This name is obscure. This adjective could apply to many things, such as a distinctive solitary tree or other solitary feature now disappeared.

Wartnaby   711231
Worcnodebie 1086 (Domesday Book)
Waercnoth (OE male personal name) + OScand *by* 'farmstead, village'.
'The settlement associated with *Waercnoth*'.
This OE personal name with the OScand. suffix *by* indicates that there was an already existing settlement here which may have been usurped by an incoming Viking.

Welby (deserted)   725210
Alebi 1086 (Domesday Book)
Ali (OScand. male personal name) + OScand. *by* 'farmstead, village'.
'The settlement associated with *Ali*'.

Welham   766925
Weleham 1086 (Domesday Book)
This could be either *Weola* (OE male personal name) or it contains an OE form *wella* of the river Welland + OE *ham* 'farmstead, village'. The most likely meaning is 'the settlement associated with *Weola*'. The element *ham* occurs in places that were settled early in the Anglo-Saxon period.

Wellsborough (deserted)   635024
Wethelesberg 1181 (Nichols)
OE *hweowol* 'wheel' + OE *beorg* 'hill'.
'The curving hill'.

Westcotes (Leicester)   576037
Westcote 1205 (Dugdale)
'The western cottage(s)'. This lay in the old West Field of Leicester and has given its name to the Westcotes district of Leicester.

West Langton   720915
Langestone 1086 (Domesday Book)
(See Church Langton)

Whatborough (deserted)   767060
Wetburga 1086 (Domesday Book)
OE *hwaete* 'wheat' + OE *beorg* 'hill'.
'The hill where wheat is grown'.

Whetstone 557935

Westham 1086 (Domesday Book)
OE **hwet-stan** 'a whetstone'.

'The whetstone'. The meaning of this name is unclear, it might be a reference to a standing stone (now gone) or it may refer to outcrops of rock which were suitable for use as whetstones. Fine-grained syenite out-crops here.

Whittington (deserted) 486083

Withington 1209 (*Placitorum Abbreviato*)
Hwita (OE male personal name) + OE **tun** 'farmstead, village, small estate'.

'The settlement associated with **Hwita**'.

Whitwick 438157

Witewic 1086 (Domesday Book)
Hwita (OE male personal name) or OE **hwit** 'white'+ OE **wic** 'farm, usually a dairy farm'.

There are two equally plausible interpretations of this name. It could be 'the dairy farm associated with **Hwita**' or 'the white dairy farm'. There is an outcrop of white sandstone nearby which makes this interpretation a real possibility. The early spellings for **Hwita** and **hwit** are the same.

Wigston Magna 598986

Wichingestone 1086 (Domesday Book)
Vikingr (OE male personal name) + OE **tun** 'farmstead, village, small estate'.

'The settlement associated with **Vikingr**'. It is possible that this name could mean 'the Vikings' settlement' but this is unlikely in an area of Viking settlement when many villages would have been either founded by or taken over by the Scandinavian incomers. Magna began to be added to the name in the sixteenth century.

Wigston Parva 465898

Wicgestane 1086 (Domesday Book)
Wicg (OE male personal name) + OE **stan** 'stone'.

'The stone associated with **Wicg**'. The stone might have been either a standing stone, or, more likely a Roman mile-post or gravestone. The Roman Watling Street forms a boundary with Wigston Parva which is only half a mile from *Venonae*, High Cross, the major crossroads of Roman England where the Fosse Way crossed Watling Street.

Willesley (deserted)   340146

Wivelesleie 1086 (Domesday Book)
Wifel (OE male personal name) + OE *leah* 'wood, clearing in a wood'.
'The woodland clearing associated with *Wifel*'.

Willoughby Waterleys   575925

Wilcheebi 1086 (Domesday Book)
OE *wilig* 'willow(s) + OScand. *by* 'farmstead, village'.
'The settlement among or by the willows'. *Waterles* appears for the first time in 1420, the meaning is water-leas 'water-meadows'. This is a very watery parish which lies on gravel and clay with the Whetstone Brook running through it. Waterleys would have been addded to distinguish this village from Willoughby on the Wolds.

Willowes (deserted)   660180

Wilges 1086 (Domesday Book)
OE *wilig* 'willows'.
'The willows'.

Wilson   405247

Wyueleston 1203 (*Curia Regis* Rolls)
Wifel (OE male personal name) + OE **tun** 'farmstead, village, small estate'.
'The settlement associated with *Wifel*'. In nearby Breedon-on-the-Hill a *Wifeles Thorpe* (now lost) is recorded in 972, this is likely to be the same *Wifel* as in Wilson.

Wistow (deserted)   644968

Wistanestou 1086 (Domesday Book)
Wigstan (OE male personal name) + OE *stow* 'a holy place'.
'The holy place of *Wigstan*'. Legend says that *Wigstan*, an Anglo-Saxon royal prince, was murdered here in AD 849 by his cousin *Beortwulf* who wanted the throne for himself. *Wigstan* was declared a saint and martyr and the spot where he died became a sacred place of pilgrimage. Wistow church is reputed to have been built on the site of the assassination.

Withcote (deserted)   797059

Wicoc 1086 (Domesday Book)

OE *withig* 'withy, willows' + OE *cocc* 'a hillock'.

'The hillock(s) where willows grow'. This must refer to an unusual profusion of willows at this place.

Witherley   326975

Withredele 1086 (Domesday Book)

Wigthrtyth (OE female personal name) + OE *leah* 'woodland, woodland glade'.

'The woodland glade associated with *Wigthrtyrh*'. This is one of only two instances in Leicestershire which record a woman's name; the other is Great Bowden.

Woodcote (deserted)   354187

Vdecote 1086 (Domesday Book)

OE *wudu* 'wood' + OE *cot* 'cottage'.

'The cottage in the wood'.

This settlement disappears from the records by the mid-fourteenth century, it may always have been just a few cottages, barely even a hamlet.

Woodhouse   539155

les Wodehouses 1209–35 (Episcopal Registers)

OE *wudu* 'wood' + OE *hus* 'house'.

'The house(s) in the wood'.

Woodhouse Eaves   530143

les eves 1481 (Court Rolls)

(See Woodhouse.)

OE *efes* 'the edge of the wood'.

Woodthorpe   544175

Torp 1236 (Book of Fees)

OScand. *thorp* 'outlying, dependent, farmstead, village'.

This small settlement would have been settled from Loughborough. Wood was added in 1284, to distinguish this place from the many thorps in the area.

Worthington   408205

Werditone 1086 (Domesday Book)

Worth (OE personal name) + **ing-tun** 'farmstead, village, small estate'.

'The settlement associated with **Worth**'.

Wycomb   775249
Wiche 1086 (Domesday Book)

OE **wic** + OE **ham**

This is the compound OE **wic-ham** meaning 'an Anglo-Saxon settlement at or near a former Roman small town'. The element **wic** reflects the Latin **vicus** meaning small town. Wycomb lies less than a mile from the important Roman site at Goadby Marwood where many Roman artefacts and burials have been found.

Wyfordby (deserted)   792189
OE **wig** the literal meaning of this word is 'an idol' and this might refer to a pagan shrine. **Wig** could also mean 'a battle or a military force' + OScand. **by** 'farmstead, village'.

'The settlement by the ford where there is a shrine' or 'a holy place possibly associated with an important battle'. Both interpretations suggest that this was a place of religious significance in antiquity.

Wykin   405953
Wicha 1169 (Danelaw Charters)

OE **wic** 'dwelling, building, a collection of buildings for a specific purpose, hamlet'.

'A hamlet, a dairy farm'.

Wymeswold   603234
Wimundeswalde 1086 (Domesday Book)

Wigmund (OE male personal name) + OE **wald** 'woodland, tract of woodland, high forest land'.

'High tract of woodland associated with **Wigmund**'. The name **Wigmund** also occurs at Wymondham but there does not appear to be a connection between these two places

Wymondham   853187
Wimundesham 1086 (Domesday Book)

Wigmund (OE male personal name) + OE **ham** 'homestead, village, estate'.

'The place associated with **Wigmund**'. This is the same personal name as Wymeswold which lies 15 miles to the west; there is no detectable connection between these two places.

STREET NAMES WHICH SURVIVE FROM MEDIEVAL LEICESTER

Many street names in Leicester are ancient although none of them are as early as most of the village names. Almost all of these old names are found within the walls of the medieval town.

Abbeygate   1312 *Labeigate*
OScand. *gata* 'road'.
The road to the Augustinian Abbey of St Mary *de Pratis* which was founded in the twelfth century. It lay to the north of the town.

Applegate   1284 *Appellane*, 1457 *Appylgate*
'The road (OScand. *gata*) among the apple trees'.
There would have been orchards here in medieval times.

Belgrave Gate   1275 *Belegrauegate*
OScand. *gata* 'road' that led to Belgrave. Until this road was re-routed in the middle ages it formed part of the Roman Fosse Way, the long-distance route-way from Exeter to Lincoln.

Black Friars Street   1484 *Blac Freres* Lane
This road marks the site of the Dominican Friary. The Dominicans wore black habits.

Braunstone Gate   1381 *Braunstonwey*
OE *wey* 'road, way' + OScand. *gata* 'road' to Braunstone.

Cank Street   1587 *The Cank Street*
OE canc 'a small rounded hill'.
No evidence of this hill remains today.

Church Gate   1464 *Sent Margett lane*
'The lane that leads to St Margaret's Church'.

Fosse Way   1381 *Fosse* OE *foss* 'a ditch'
This name records the major Roman road which ran from Lincoln to Exeter. The road entered the town at West Bridge and Braunstone Gate and left through the eastern gateway. From the medieval period

onwards the Fosse Way by-passed the city to the west of Leicester Abbey crossing the River Soar south of Birstall.

Friar Lane   1391 *le Frere* Lane

The site of the friary of the Franciscan monks. They were known as the grey friars because of the colour of their habit.

Gallowtree Gate   1290 *le Galtregate*

OE **galg-treow** 'the gallows tree' + OScand. **gata** 'the road leading to the gallows tree'.

The gallows were sited outside the eastern corner of the town.

High Cross Street   1392

Latin **altam crucem** 'high cross'.

A weekly market was once held here.

High Street   1253

Latin **altam stratum** 'the chief or high street'.

The medieval High Street was High Cross Street which ran between the north and south gates. With the eastward shift of the commercial centre of the town the street known today as the High Street became the new main street. It was first recorded in 1523 as *'the high stret which is in the Est Yate'*. Its earlier name was the *Swines Market*.

Holy Bones   1349 *le Holybonse*

This name probably records a medieval discovery of unknown graves outside the church yard of St Nicholas Church. It runs next to the excavated remains of the Roman baths which were constructed above an earlier Iron Age settlement. An Anglo-Saxon church may have been sited on the bath site near to the church. The 'holy bones' could date from any of these periods of occupation.

Horsefair Street   1546 *horsefayre lane*

'The lane where the horse market was held'. Town Hall Square now stands on this site.

Humberstone Gate   1286 *Humbirstongategate*

OScand. **gata** 'road' which led to Humberstone.

Jewry Wall   1665 *the Jury Wall*

There is doubt about the origin and meaning of this name. It might refer to the Jews' quarter of the town. There is no direct evidence for a Jewish quarter although there is a charter of 1250 that decrees that quarter which would also have been the case in Leicester. Another possible interpretation of the name is that it might be a corruption of *Jurats*, the 24 medieval town councillors whose meetings might have been held in or close by to St Nicholas Church. The wall itself is part of the Roman baths.

Loseby Lane   1488 *Loseby lane*
Named from Henry *de Louseby* who held land in the parish of St Martin c.1300.

Millstone Lane   1453 *Mylston Lane*
'Lane paved with millstones'.

Sanvey Gate   1392 *Senueygate*
OE **sand-weg** 'the sandy way' + OScand. **gata** 'road'.

In 1316 this road was also known as **le Sckeyth** a name that derives from the ON **sketh** 'a track, a race, a race-course'. There is no evidence of a racecourse here although Vikings were passionate horse-racers which would make the racecourse interpretation a plausible suggestion. In 918 the Anglo-Saxon Chronicle records the Viking army as being centred on Leicester maybe they were horse-racers who gave this street its name. The Roman walls and banks, still a prominent feature, would have provided an excellent raised area from which to watch the races.

The Newarke   1361 *le Newerk*
OE **niwe** 'new' + OE **werke** 'fortifications, buildings'.

This 'new work' was the walled enclosure with towers and a gatehouse, built as an extension of the castle. In 1330 Henry, Earl of Lancaster and Leicester, founded, a hospital for fifty poor, sick people within these walls. As Trinity Hospital this foundation survived into the nineteenth century.

RUTLAND

Rotelande 1053 (Anglo-Saxon Writs)
Roteland 1086 (Domesday Book)
Rota (OE male personal name) + OE *land* 'land, domain'
'Rota's land' or 'the land belonging to Rota'

In place-names the use of the word *land* usually refers to a tract of land of considerable extent, as in a county or country but it can also mean a piece of land as small as a field. In the case of Rutland *land* refers to a small kingdom the boundaries of which may go back into the Iron Age or even earlier.

Nothing definite is known about Rota, only his name, used in the place-name Rutland, appears in the records. However it is possible to make a few informed guesses about him.

Rota would have been connected with the early Anglo-Saxon settlement of Rutland, no later than the sixth or early seventh centuries. He was probably a minor king among the loose federation of Anglo-Saxon tribal groups known as the Middle Angles, who settled in this part of England. The Middle Angles do not appear to have had one dominating dynasty as was the case in many other areas of England at this time. Rota would have been a man of considerable power and importance for his name to have become attached to such a large tract of land. No other unit of land of such size in England is named after one man. Rota would have been buried within his own kingdom but his burial place has not been found.

HUNDREDS OF RUTLAND

Domesday Book records *Roteland* as consisting of two hundreds, Alstoe and Martinsley. The remainder of what is now the present county of Rutland was listed under Northamptonshire. This area comprised the Hundred of Witchley which, by 1166, was divided into East Hundred and Wrangdike Hundred. At Domesday, Rutland was connected with Nottinghamshire with Alstoe Hundred divided between Thurgarton Wapentake (Hundred) and partly in Broxtow Wapentake (Hundred).

Alstoe Hundred
Alfnodestou 1086 (Domesday Book)

Aelfnoth (OE male personal name) + OE **stow** 'place' or 'holy place'. 'The holy place associated with **Aelfnoth**'.

Alstoe Hundred lay in the north-west part of Rutland. The meeting place was probably near the present day Alstoe House north-east of Burley. Earthworks, which were probably connected with the Hundred meeting, can still be seen.

Martinsley Hundred

Martineslei 1086 (Domesday Book)

Martin (OE male personal name) OE **leah** 'wood, woodland clearing, woodland glade'.

'The woodland associated with **Martin**'.

Martinsley Hundred roughly comprises the low land of central Rutland. The Hundred meeting place was probably at Martinsthorpe, south of Oakham.

Witchley Hundred

Hwicceslea east hundred & Hwicceslea west hundred 1075 (Geld Roll, Northants.)

Wiceslea 1086 (Domesday Book)

OE **Hwicca-leah** 'the woodland of the **Hwicce**'.

Witchley is the name of two hundreds in the Northamptonshire Geld Roll of 1075. By 1086 these two hundreds are treated as one single hundred. The name records the early Anglo-Saxon tribal folk-group the **Hwicce** whose name is also recorded in the place-name Whissendine. Witcherley Farm was probably the Hundred meeting place.

Wrangdike Hundred

Wrangediche 1166 (Pipe Rolls)

OE **wrang** 'crooked' OE **dic** 'ditch'.

This is **Hwicceslea west hundred** first recorded in 1175 (see Witchley Hundred).

Wrangdike Hundred lay in the south of Rutland between Eye Book and Tixover with the River Welland forming the boundary with Northamptonshire. The site of the meeting place may have been in Seaton where the name **syrepol** is recorded, this could be OE **scir** 'shire' + OE **pol** 'pool'. Wrangdike has always been styled as a Hundred, never as a Wapentake.

Oakham Soke Hundred

Hundreda de Okeham cum Martynley 1428 (Feudal Aids)

Oakham Soke Hundred originally formed part of Martinsley Hundred. It was first recorded in 1428 and later appeared 1610 on Speed's map of Rutland. The territory of this hundred comprised the manors and townships held by the castle and manor of Oakham.

RUTLAND SETTLEMENTS

Alstoe (deserted)   895119

Alfnodestou 1086 (Domesday Book)

Aeflnoth (OE male personal name) + OE s*tow* 'place of assembly, a holy place'.

'The assembly/meeting place associated with *Aelfnoth*'.

Alstoe Hundred/Wapentake moot site (meeting place) can be seen as a pronounced earthwork beside Alstoe House in Burley parish.

Alsthorpe (deserted)   883103

Alestanestorp 1086 (Domesday Book)

Alhstan (OE male personal name) + OScand. *thorp* 'outlying, dependent farmstead or village'.

'The outlying settlement associated with *Alhstan*'. This place lay in what is now the Burley estate.

Ashwell   866137

Exewelle 1086 (Domesday Book)

OE *aesh* 'ash tree(s)' + OE *wella* 'spring or stream'.

'The well or stream where the ash trees grow'.

Ayston   859009

Aethelstanestun 1046 (*Codex Diplomaticus Aevi Saxonici*)

Aethelstan (OE male personal name) + OE *tun* 'farmstead, village, small estate'.

'The settlement associated with *Aethelstan*'. *Aethelstan* is named in a charter of 1046 (the only one of its kind in Leicestershire or Rutland) as the holder of an estate the territory of which now comprises the parish of Ayston. The survival of the charter allows us a rare glimpse of a name being coined. Before this date the estate was known as *Thornham*. Most early place-names are lost forever as they were not recorded before changing circumstances brought about a name change, as was the case here at Ayston.

Barleythorpe 848097

Thorp juxta Ocham 1200 (Westminster Domesday Book)
Bolaresthorp 1203 (Feet of Fines)

OScand. ***thorp*** 'outlying dependent settlement'. *John le Bolar* is mentioned in connection with Oakham in 1200. The name developed into Barleythorpe as *Bolar* became meaningless and, through the association of sound, barley replaced the family name.

Barnsdale 904091

Bernardeshull 1202 (Assize Rolls)
Beornheard (OE male personal name) + OE ***hyll*** 'hill'.
'The hill associated with ***Beornheard***'.

The present day name evolved when the personal name ***Beornheard*** ceased to have any meaning with ***deshull*** becoming 'dale'.

Barrow 891152

Berc 1206 (*Curia Regis* Rolls)
OE ***beorg*** 'hill, mound, burial mound'.

This element often refers to an artificial mound such as a prehistoric burial mound or barrow. The village stands on a steep hill on the top of which there is a burial mound which would have given Barrow its name.

Barrowden 945999

Berchedone 1086 (Domesday Book)
OE ***beorg*** 'hill, mound, burial mound' + OE ***dun*** 'hill, open hill country'.
'Hill on which there are prehistoric burial mounds'.

The burial mounds which would have been old before the Anglo-Saxons settled here, may have been reused by the Anglo-Saxons.

Beaumont Chase

Bellomonte 1203 (Patent Rolls)
OFr. ***beau-mont*** 'beautiful hill' + OFr. ***chace*** 'a tract of ground for breeding and hunting wild animals'.

This is the only major name in Rutland which dates from Norman Conquest in 1066.

Belmesthorpe 043102
Beolmesthorpe 927 (*Codex Diplimaticuc Aevi Saxonici*)
Belmestorp 1086 (Domesday Book)
Beornhelm (OE male personal name) + OScand. *thorp* 'outlying, dependent settlement'.
'The settlement associated with *Beornhelm*'

Belton 816014
Beltone 1205 (Dugdale, *Monasticon Anglicanum*)
OE *bel* 'fire, funeral pyre, beacon, interval, a clearing' + OE *tun* 'farmstead, village, small estate'.
'A settlement where there is a beacon'. Belton is on the border of Leicestershire and Rutland so this might have been a place where beacon fires were lit. It also lay in wooded country so the meaning could also be 'a woodland clearing'. The balance of probabilities lies with it being a beacon or look-out.

Bisbrooke 887996
Bitlesbroch 1086 (Domesday Book)
Bitel (OE male personal name) + OE *broc* 'brook, stream'.
'The brook/stream associated with *Bitel*'.

Braunston 833006
Brantestone 1167 (Pipe Rolls)
Brant (OE male personal name) + OE *tun* 'farmstead, village, small estate'.
'The settlement associated with *Brant*'.

Brooke 850057
Broc 1176 (Pipe Rolls)
OE *broc* 'the brook, stream'.
'The brook' (upper reaches of the Gwash).

Burley-on-the-Hill 883103
Burgelai 1086 (Domesday Book)
OE *burh* 'fortified place + OE *leah* 'wood, woodland clearing, glade'.
'Clearing in the wood near the fortified place'. The fortified place was almost certainly Hambleton which was the central place of an early Anglo-Saxon kingdom.

Caldecott 868937

Caldecot 1246 (Charter Rolls)

OE *cald* 'cold, exposed' + OE *cot* 'cottage, shelter'.

This could refer to a shelter for livestock in a cold or exposed place. It might also refer to a shelter for travellers on the Roman road which crosses the Welland just to the south of the village. Many places with the name Caldecott are found on major routes and Roman roads making the interpretation of a shelter for travellers the more likely of the two.

Clipsham 970164

Kilpesham 1203 (Feet of Fines)

Cylp (OE male personal name) + OE *ham* 'farmstead, village, estate'.

'The settlement associated with *Cylp*'.

Although this name does not appear in the records until the early years of the thirteenth century this settlement almost certainly dates from the early years of the Anglo-Saxon settlement of the area.

Cottesmore 903136

Cottesmore (Anglo-Saxon will)

Cott (OE male personal name) + OE *mor* 'moor, wasteland'.

'A stretch of moorland associated with *Cott*'.

Edith Weston 927054

Westone 1167 (Documents preserved in France)

OE *west* 'west, western' + OE *tun* 'farmstead, village, small estate'.

'The western estate'. This place lies to the west of the important Anglo-Saxon royal settlement of Ketton. The name Edith is associated with Queen *Eadgyth*, wife of Edward the Confessor, who had holdings in Rutland at the time of the Norman Conquest.

The affix *Edy* is first recorded in 1263.

Egleton 876075

Egoluestun 1218 (Pleas of the Forest)

Ecgwulf (OE male personal name) + OE *tun* 'farmstead, village, small estate'.

'The settlement associated with *Ecgwulf*'.

Empingham 950085

Epingeham 1086 (Domesday Book)

Empa (OE male personal name) + OE ***ham*** 'village, homestead, estate'.

'The settlement associated with the people of ***Empa***'. The habitative element ***ham*** occurs early in the Anglo-Saxon period and ***Empa's*** people or folk would probably have constituted a 'kin' group whose central place, or 'home' would have been where Empingham now lies. The word 'kin' in Anglo-Saxon society usually carries a wider meaning than that of a group united by ties of blood or marriage, kin might also include people from the same geographical area who could have been regarded as 'one of us'.

Essendine   049128
Esindone 1086 (Domesday Book)
Essa (OE male personal name) + OE ***denu*** 'valley'.
'The valley associated with ***Essa***'.

Exton   920112
Exentune 1086 (Domesday Book)
OE ***oxa*** 'ox(en)' + OE ***tun*** 'farmstead, village, small estate'.
'The ox farm' or 'the place where the oxen were kept'.

Flitteris Park   810070
Flyteris 1252 (patent Rolls)
OE ***flyt*** 'strife, dispute' + OE ***hris*** 'brushwood'.
'Area of brushwood-covered land of disputed ownership'.

This land lay on the boundary between Leicestershire and Rutland in the parishes of Whissendine and Teigh. The period of this dispute is unclear but it would have predated the time of Domesday and almost certainly refers to disputes connected with the boundary between the two counties.

Fregsthorpe (deserted)   001049
Frygisthorpe 1300 (Dean and Chapter MSS)
Frithegist (OScand. male personal name) + OScand. ***thorp*** 'outlying dependent settlement'.
'The settlement associated with ***Frithegist***'.

A man with the name of ***Fredgis*** is recorded in the Domesday entry for neighbouring Empingham as holding land at the time of Edward the Confessor. This may be the same man who gave his name to Fregsthorpe.

Geeston (deserted) 987040

Gyston 1286 (Assize Records)

Gyssa (OE male personal name) + OE ***tun*** 'farmstead, village, small estate'.

'The settlement associated with ***Gyssa***'.

The earthworks associated with this 'lost' settlement can still be seen.

Glaston 896005

Galdestone 1086 (Domesday Book)

OScand. ***Glathr*** (personal name) + OE ***tun*** 'farmstead, village, small estate'.

Glathr is a Norwegian name. Along with nearby Normanton 'the settlement of the Norwegians' these two names would seem to indicate the presence of Norwegians in the area. Rutland, unlike many parts of neighbouring Lincolnshire and Leicestershire, has few place-names of Scandinavian origin.

Great Casterton 002088

Castretone 1086 (Domesday)

OE ***ceaster*** 'fortified Roman town' + OE ***tun*** 'farmstead, village, small estate.

'The settlement at the fortified Roman town'.

When the Anglo-Saxons identified a site as that of a Roman fortified place they usually used the element ***ceaster***. 'Great' distinguishes this place from Little Casterton. An early Roman fort was established here with a town growing up beside it where the Roman Ermine Street crossed the River Gwash. The Romano-British name appears to have fallen out of use before the arrival of the Anglo-Saxons.

Greetham 924146

Gretham 1086 (Domesday Bok)

OE ***greot*** 'gravel' + OE ***ham*** 'homestead, village, estate'.

'The settlement on gravely stoney soil'. Northampton Sands ironstone outcrops here.

Gunthorpe (deserted) 870057

Gunetorp 1200 (*Placitorum Abbreviato*)

Gunni (OScand. personal name) + OScand. ***thorp*** 'outlying, dependent farmstead or village'.

'The farmstead or village associated with ***Gunni***'.

Hambleton   900076

Hamelduna 1067 (*Westminster Domesday Book*)
Hameldune 1086 (Domesday Book)
OE *hamol* 'maimed, crooked, mutilated' + OE *dun* 'hill, a flat-topped hill, expanse of open hill country'.
'The settlement on the crooked/maimed hill in open country'.
This name refers to the distinctive shape of the hill here. In the early period of Anglo-Saxon settlement Hambleton was the administrative centre of Rutland, a function which by the time of Domesday had been taken over by Oakham.

The original Anglo-Saxon settlement was at Upper Hambelton; Middle and Lower Hambleton were destroyed in 1975 when Rutland Water was constructed.

Hardwick (deserted)   950085

Herdewic 1281 (Calendar of Close Rolls)
OE *heorde-wic* 'the (sheep) herd farm'.

Horn (deserted)   956119

Hornan 852 (*Cartularium Saxonicum*)
Horne 1086 (Domesday Book)
OE *horn* 'a horn'.
This name must refer to the sharp bend here in the North Brook. No village survives but the earthworks can clearly be seen including a well-defined moated manor house site.

Ingthorpe (deserted)   998089

Ingelthorpe 1189 (Calendar of Charter Rolls)
Torp 1125 (*Chronicon Petroburgense*)
Ingi (OScand. male personal name) + OScand. *thorp* 'outlying, dependent farmstead or village'.
'The settlement associated with *Ingi*'.

Ketton   982043

Chetene 1086 (Domesday Book)
This is an old name of the River Chater. Celtic *ceto* 'a wood' + OE *ea* 'river'. The whole word is probably derived from the Celtic word *ceto-dubron* meaning 'a forest stream'.

Kilthorpe (deserted) 985033

Ketelistorp 1250 (Court Rolls)

Ketil (OScand. male personal name) + OScand. ***thorp*** 'secondary, dependent farmstead or village'.

'The settlement associated with ***Ketil***'.

Langham 844113

Langeham 1202 (Dugdale *Monasticon Anglicanum*)

The second element might be OE ***ham*** 'homestead, village, estate' or OE ***hamm*** 'meadow, water meadow' + OE ***lang*** 'long'. Distinguishing between the elements ***ham*** and ***hamm*** is difficult as the early spellings of this place-name allow for either interpretation. The topography would suggest that this is a ***hamm***.

Leighfield 8204

Leghe 1266 (Forest Charters)

The royal forest of Leighfield once covered all of Rutland it was later limited to the south west of the county. OE ***leah*** is used here in its early sense of 'woodland' and OE ***feld*** 'open country' was added when the forest began to be cleared.

Little Casterton 018099

(See Great Casterton)

Lyddington 877970

Lidentone 1086 (Domesday Book)

Hlyda (OE male personal name) + OE ***ingtun*** 'farmstead, village, small estate'.

'The settlement or small estate associated with ***Hlyda***'.

Lyndon 907044

Lindon 1167 (Pipe Rolls)

+ OE ***lind*** 'lime-tree' + OE ***dun*** 'hill, a flat-topped hill, expanse of open hill country'.

'Open hill country covered with lime trees'.

Manton 881047

Manatone 1130 (Documents preserved in France)

Manna (OE male personal name) + OE ***tun*** 'farmstead, village, small estate'.

'The settlement associated with ***Manna***'.

Market Overton   885165

Overtune 1086 (Domesday Book)
OE *ofer* 'bank, river bank, ridge' + OE *tun* 'farmstead, village, small estate'.
'The settlement on or near a ridge'. An early Roman fort was established here the church now stands on the site. A market was recorded here as early as 1200.

Martinsthorpe (deserted)   866045

Martinestorp 1206 (Curia Regis Rolls)
OE **Martin** (OE personal name developed from the Latin **Martinus**) + OScand. *thorp* 'outlying dependent farmstead, village'.
'The settlement associated with Martin'.

Martinsley (lost)   867046

Martensley 1536 (Mss of the Earl of Denbeigh)
'The woodland clearing associated with **Martin**'.
This was probably the moot site (meeting place) for Martinsley Hundred.

Morcott   924007

Morcote 1086 (Domesday Book)
OE *mor* 'moor or wasteland' + OE *cot* 'cottage, shelter (often for animals)'.
'The cottage or animal shelter on moorland'.

Nether Hambleton (deserted)

(See Hambleton.)

Normanton (deserted)   933064

Normantone 1183 (Pipe Rolls)
OE **Northman** 'Norwegian' + OE *tun* 'farmstead, village, small estate'.
'The settlement of the Norwegians'.

This name reflects the ethnic differences of the Viking invaders/settlers. Nearby Glaston also embodies a Norwegian personal name. The village was depopulated in 1764 when Sir Gilbert Heathcote moved the inhabitants to Empingham so that he could lay out his park. His house was later dismantled and rebuilt in Scraptoft village, Leicestershire. It is now a restaurant.

North Luffenham   934033
Lufenham 1086 (Domesday Book)
Luffa (OE male personal name) + OE *ham* 'homestead, village, estate'.
'The settlement associated with *Luffa*'.
The element *ham* is an indicator of early Anglo-Saxon settlement. Excavations here have produced many high status obects, incuding several swords dating from the sixth and seventh centuries.

Oakham   861089
Ocheham 1086 (Domesday Book)
Occa (OE male personal name) + OE *ham* 'village or homestead'.
'The village or estate associated with *Occa*' or '*Occa's* village or estate'.
By the time of Domesday Oakham had emerged as the the most important settlement in the county having taken over from Hambleton which had been pre-eminent in the early Anglo-Saxon period.

Pickworth   993138
Pichewutha 1228 (*Curia Regis* Rolls)
Pica (OE male personal name) + OE *worth* 'enclosure, settlement'.
'The farmstead or settlement associated with *Pica*'.

Pilton   915029
Pilton 1202 (Assize Rolls)
OE *pyll* 'pool, creek in a river' + OE *tun* 'farmstead, village, small estate'.
'Settlement at or near a creek in the river'.

Preston   870024
Prestetona 1130 (pipe Rolls)
OE *preosta* 'priest' + OE *tun* 'farmstead, village, small estate'.
'The estate of the priest(s)'. This land would have been endowed for the clergy.

Ridlington   847027
Redlinctune 1086 (Domesday Book)
Redel (OE male personal name) + OE *ingtun* 'farmstead, village, small estate'.
'The settlement associated with *Redel*'. The form *ingtun*, which belongs to the eighth and ninth centuries, probably indicates

possession usually by a noble, who was not necessarily resident. Ridlington was an Anglo-Saxon royal possession in Domesday and part of the royal lands centred on Hambleton.

Ryhall   036108

Righale 1042 *(Codex Diplomaticus Aevi Saxononici)*
Riehale 1086 (Domesday Book)
OE **halh** 'nook, corner of land' + OE **ryge** 'rye'.
'A corner of land where rye is grown'. The corner could be the bend in the River Gwash here.

Sculthorpe (deserted)   934033

Sculetorp 1086 (Domesday Book)
Skuli (OScand. male personal name) + OScand **thorp** 'secondary, dependent farmstead or village'.
'The settlement associated with **Skuli**'.

Seaton   904983

Segentone 1086 (Domesday Book)
Seaga (OE male personal name) + OE **tun** 'farmstead, village, small estate'.
'The settlement associated with **Seaga**'.

Snelston (deserted)   867953

Smelistone (Domesday Book)
Snell (OE male personal name) + OE **tun** 'farmstead, village, small estate'.
'The settlement associated with **Snell**'.

South Luffenham   941019

(see North Luffenham)

Stoke Dry   856967

Stoche 1086 (Domesday Book)
This name could be either OE **stoc** 'place, religious place', or OE **stocc** 'tree-trunk(s)/stump(s), 'a standing place for cows or where cows were milked', 'a dairy farm'.

The latter interpretation is the more likely here. **Stoc** and **stocc** in place-names are often unclear and inconsistent. 'Dry' refers to the land here being dry in contrast with what would have been the marshy conditions lower down the hill near the Eye Brook.

Stretton   949157

Stratone 1086 (Domesday Book)

OE ***straet*** 'Roman road' + OE ***tun*** 'farmstead, village, small estate'.

The settlement which lies on or near the Roman road'. Stretton lies on the Roman Ermine Street.

Teigh   865160

Tie 1086 (Domesday Book)

OE ***teag*** 'small enclosure'.

In 1449 there is a documentary reference here to *burh-steal medewe* 'the meadow where there was a fortified place'. The ***teag*** might have been an enclosure attached to this stronghold.

Thistleton   913180

Tisteltune 1086 (Domesday Book)

OE ***thistel*** 'thistle' + OE ***tun*** 'farmstead, village, small estate'.

'The settlement where thistles abounded'. Thistles grow abundantly in places where buildings once stood as they thrive on the phosphates produced by human activity. Thistleton was an important Romano-British market town in which there was a Roman temple dedicated to the Celtic god *Veteris*. The Roman name of this town fell out of use before the Anglo-Saxons arrived.

Thorpe-by-Water   894965

Torp 1086 (Domesday Book)

Thorpbythewatir 1428 (Feet of Fines)

OScand. ***thorp*** 'out-lying/dependent settlement'. The 'water' is the River Welland.

Tickencote   990095

Tichecote 1086 (Domesday Book)

OE ***ticcen*** 'young goats' + OE ***cot*** 'cottage, shelter usually for animals'

'The shelter for young goats'.

Tinwell   006063

Tedinwelle 1086 (Domesday Book)

Tida (OE male personal name) + OE ***wella*** 'spring or stream'.

'The spring or stream associated with ***Tida's*** people

This is a 'folk' name which dates back the time of the earliest Anglo-

Saxon settlement when people would still have been coming into the country and settling initially in 'folk' or 'kin' groups.

Tixover (deserted)   971998
Tichesovre 1086 (Domesday Book)
OE *ticcen* young goats' + OE *ofer* 'bank, river bank, ridge'.
'The ridge where young goats were grazed'.

Tolethorpe (deserted)   023103
Toltorp 1086 (Domesday Book)
Toli (OScand. male personal name) + OScand. *thorp* 'out-lying/dependent settlement'.
'The outlying settlement associated with *Toli*'.

Uppingham   866996
Yppingeham 1067 (British Museum Charters)
OE *Yppingas* or *Uppingas* 'the folk/people of the uplands' + OE *ham* 'homestead, village, small estate'.
'The settlement of the people of the upland' or 'the folk who live on the hill'.
Uppingham is sited on a high ridge.

Vale of Catmose   805009
Catmose 1610 (Speed, map of Rutland)
There are no early forms of this name but it is reasonable to assume that it derives from OE *cat* 'wild cat' + OE *mos* 'bog or swamp'.
'Boggy land infested with wild cats'.

Wardley   832003
Werlea 1067 (British Museum Charters)
OE *weard* 'watch, ward' + OE *leah* 'wood, clearing in a wood'. Wardley stands on a hill above the Eye Brook, it would have made an excellent look-out. It lies next to Belton the name of which is probably means a beacon. Place-names which contain the element *weard* are often found on important boundaries. This Wardley is on the boundary between Leicestershire and Rutland.

Wenton (deserted)   891143
Weneton 1200 (Curia Regis Rolls)
Wenna (OE male personal name) + OE *tun* 'farmstead, village, small estate'.
'The settlement associated with *Wenna*'.

Whissendine   833143
Wichingedene 1086 (Domesday Book)
OE *Hwiccingas* + OE *denu* 'valley'.
This is probably an Anglo-Saxon folk name, 'the valley of the folk known as the *Hwiccingas*'. Less likely but possible *Hwicce* could be a male personal name.

Whitwell   924087
Witewelle 1086 (Domesday Book)
OE *hwit* 'white' + OE *wella* 'spring or stream'.
A small stream rises here. The reference to white probably refers to the colour of the water or the stream bed.

Wing   894029
Wengeford 1046 (Anglo-Saxon Charter)
OScand. *vengi* 'an in-field' or 'garden' which later developed into a village.

Witchley Warren (deserted)   958054
Hwiccleslea 1046 (Anglo Saxon charter)
Hwicce (male personal name) + OE *leah* 'wood, clearing in a wood'.
'Woodland belonging to the people associated with *Hwicce*.'
Hwicce is a tribal name which dates back to the earliest period of Anglo-Saxon settlement in the county. The *Hwicce* were probably the folk group who gave their name to the county of Worcestershire, although no provable connection has established between the names.

RIVERS OF LEICESTERSHIRE AND RUTLAND

Chater

Cahtere 1263

The name probably derives from the Celtic word **ceto** 'a forest stream'.

Rises east of Halstead and joins the River Welland near Stamford.

Devon

Dyuene 1252

A Celtic name which means 'deep'. In its upper part this stream runs in a ravine. It rises near Eaton and joins the River Trent at Newarke.

Eye

Eye 1545

OE *ea* 'the river'.

Rises near Saltby and flows to the Wreake at Melton Mowbray

Eye Brook

Litelhe 1218

OE *ea* 'river' + **broc** 'brook, a little river'.

Rises at Tilton and joins the River Welland at Caldecott. It forms the boundary between Leicestershire and Rutland from Allexton southwards.

Gwash

le Whasse 1263

OE **gewaesc** 'a wet place, a swamp or a marsh'.

In this case the name probably means a 'stream in a marsh'. The Anglo-Saxons often used **waesse** to denote a water course which qickly rises, floods and subsides just as quickly. It rises near Owston and Joins the River Welland near Stanford.

This word is the same as the River Weser in Germany.

Lipping

Lippinge 1218

This may be an old Anglian name which derives from the Germanic root **leib** 'to pour, to flow'.

It rises near Skeffington and joins the river Welland near Welham.

Mease
Meys 1247
OE *meo* 'bog, swamp'.
This river would have followed a swampy course; boggy land can still be seen here.
Rises near Ashby-de-la-Zouch and joins the Trent at Croxall, Staffordshire.

Rothley Brook
Hathebroc 1276
OE *roth* 'clearing or heath, OE *broc* 'brook, a little stream'.
'A brook in a clearing'.
Rises at Stanton under Bardon and joins the River Soar at Rothley.

Sence
Sceynch OE 'a drink or draught, a flow of clear drinking water'.
Rises on Bardon Hill and joins the River Anker at Atherstone, Warwickshire.

Sence
The original name of this river was *Glene* as recorded on a map of 1402, a name which may derive from the Celtic word *glano* 'holy, clean' but is more likely to come from a Celtic word meaning 'glen or valley'. The name Sence is first recorded on Speed's map of 1610 it may carry the same meaning as the River Sence above although it is unusual for two rivers so close together to have the same name.
Rises at Billesdon and joins the River Soar above Narborough.

Smite
Smite 1280
OE *smite* 'to glide' or 'slip'. 'The sliding river'.
Rises near Old Dalby and flows to its confluence with the River Selton, Nottinghamshire.

Soar
Sore 1211
This is an obscure pre-English word which may derive from the root of the Celtic word *ser* (as in the Latin *serum*) meaning 'to flow'. It is also possible that Soar is a pre-Celtic word from a non-Indo-European language the meaning of which is completely lost as nothing is known of these earlier languages.

The Soar rises south of Sharnford and flows north to join the Trent north of Ratcliffe on Soar.

Swift

Swift 1586

OE *swift* 'to move in a sweeping manner'.

The Swift rises near Lutterworth and joins the River Avon at Rugby.

Welland

Weolud 921

This name may derive from the Celtic word *vesu* 'good' with the second element deriving from another Celtic word *luaid* 'to move'. The change from *Weolud* to Welland may be the result of Scandinavian influence on the language.

Rises near Sibbertoft in Northamtonshire and joins the Wash seventy miles away.

Wreake

Wreithk 1224

OScand. *wreithk* 'twisted, crooked'.

The Wreak is a winding river running through an area of pronounced Viking influence. It rises at Waltham on the Wolds and joins the River Soar at Rothley. From its origin in Waltham on the Wolds down to Melton Mowbray the river name is Eye and from Melton onwards to the River Soar it is known as the Wreake. It is rare for a river name to change along its course. The name change in this instance must reflect the density of Viking settlement.

ELEMENTS THAT OCCUR IN LEICESTERSHIRE AND RUTLAND PLACE NAMES

abbei OFr. 'abbey'

ac OE 'oak tree'

aeppel OE 'apple', 'fruit in general'

aesc OE 'ash tree'

ald OE 'old'. Often denotes old fortifications or homesteads.

an(a) OE 'one, solitary, lonely'

anliep OE 'single, solitary, lonely'

anstiga OE 'a narrow footpath usually going up a slope'

baerlic OE 'barley'

baer-tun OE 'barley-farm'

ban OE 'bone'

bar OE 'wild or domestic boar'

beacon OE 'beacon, signal, sign'. Often marking a hill-top where beacon fires were lit for signalling.

beam OE 'tree or piece of heavy timber'

bearu OE 'a wood, a grove'

beau OFr. 'beautiful'

bece OE 'a stream, a valley'

bel OE 'a space, an interval, beacon'

beo OE 'a bee'

beorg OE ' a hill, mound, burial mound'. The Anglo-Saxons used the term **beorg** to describe burial mound which belonged to an earlier time, **hlaw** was used for their own (pagan) burial mounds.

bere OE 'barley'

blaec OE 'black, dark, dark-coloured'

bold OE 'a building'

botl OE 'a dwelling house'. This element sometimes indicates a mansion/palace.

brad OE 'broad, spacious'

bre Celtic 'hill'

broc OE 'a brook, stream'

burh OE 'a fortified place'

burh-stall OE 'site of a fortified place'

burna OE 'stream, spring'

butt OE 'a mound, stump, an archery butt'

by OScand. 'farmstead, village'

bytme OE 'head of a valley'

cald OE 'cold, bleak, exposed'

carn Celtic 'heap of stones, a cairn'

castel(l) OFrench 'castle, camp'

cat(t) OE ' a cat, wild cat'

ceaster OE 'an old fortification, Roman town'

ceto C 'a wood'

ceorl OE 'a churl, a lower class of free peasant'

chace Ofr 'land used for breeding and hunting wild animals'

cild OE 'child, young person'. This can mean a noble-born youth.

claeg OE 'clay, clayey soil'

claen OE 'clean, clean of weeds'

clif OE 'a cliff, bank'

cneo(w) OE 'knee used as in the sense of a bend in a road or river'

cniht OE 'youth, servant, soldier'. Often a place held by the retainer of a royal or noble person. Later a person raised by the king to a high military rank.

cocc OE 'a heap, hillock, a dunghill'

col OE 'coal, charcoal'

copped OE 'pollarded, cut down somewhat'

cot OE 'cottage, hut, shelter'. Most commonly this element indicates a humble dwelling.

craeft OE 'a machine, engine'. This possibly refers to some sort of mill (not wind).

crawe OE 'crow'

croft OE 'small enclosed field, piece of land'

cros OFr. 'a cross'

cwen OE 'queen, woman'. Examination of the context should indicate which is the most likely meaning.

cweorn OE 'quern, handmill'

cyning OE 'king'

dael OE 'valley, pit, hollow'

denu OE 'valley'

draeg OE 'drag, portage, slip-way'. The fundamental meaning is to drag/draw.

dryge OE 'dry, dried up'. This element usually refers to a dried-up river course.

dun OE 'hill, a flat-topped hill, an expanse of open hill country'

ea OE 'river, stream'

east OE 'east, eastern'. Usually meaning east of a more important or older-established place.

efes OE 'eaves, edge of wood, border'

eg OE 'an island'. This can also indicate dry land in fenny land.

eorl OE 'nobleman'

faesten OE 'stronghold'

feld OE 'open country'. The modern use of 'field' arose from the fourteenth century onwards when the large, open, common fields were enclosed.

fennig OE 'marshy, muddy'

flit OE 'strife, dispute'

ford OE 'ford, shallow place across a stream'

forest OFr. 'a large tract of woodland, a forest'

foss OE 'ditch'

fox OE 'fox'

fyrhth OE ' a wood, woodland, wooded countryside'

gall OScand. 'barren or sterile place'

gara OE 'a gore, triangular piece of land'

gata OScand. 'a way, road, path'

geat OE 'an opening, a gap'

geirr OScand. 'spear'

glennos Celtic 'a glen, valley'

gnipa OE 'steep rock, over-hanging rock in a valley'

gop OE 'slave, servant'

gos OE 'a goose'

graf OE 'grove, copse, thicket'

grange OFr. 'a granary, barn', later 'an outlying farm belonging to a religious house'.

greot OE 'gravel'

grof OScand. 'a stream, a pit, a hollow made by a stream'

gryfja OScand. 'a hole, a pit, a small deep valley'

gylden OE 'golden'

haefera OE 'oats'

haeg OE ' fence, an enclosure'

haeth OE 'heath, heather'

hafocere OE 'a hawker, falconer'

hald OE 'shelter, refuge'

halh OE 'a nook, corner of land'

halig OE 'holy, sacred'

ham OE 'village, village community, homestead'

hamm OE 'enclosure, meadow, water-meadow'

h**amol** OE 'maimed, mutilated, crooked'

har OE 'grey, hoar'. Especially grey through being overgrown with lichen, developing into the meaning of a boundary stone.

heafod OE 'head, hill, the top' (of something)

heah OE 'high'

heithr OScand. 'heath, uncultivated land'

heorde-wic OE 'herd farm'

here OE 'an army'

hid OE 'a hide of land', a notional unit of land for the support of one free family and its dependants for a year (approximately 120 acres).

hlaw OE 'mound/hill' **Hlaw** usually referred to a burial mound of the Anglo-Saxons themselves, for burial mounds of earlier times they used the term **beorg**.

hluttor OE 'clear, bright'

hlyde OE 'a noisy stream'

hoh OE 'a heel, a spur of land'

hol OE 'a hole a hollow'

holt OE 'a wood, a thicket'

horna OE 'horn' or something resembling a horn'

hris OE 'shrubs, brushwood'

hungor OE 'hunger, famine'

hwaete OE 'wheat'

hweowal OE 'wheel, something circular'

hwet-stan OE 'a whetstone'. This word usually refers to a place where rock suitable for whetstones was found.

hwit OE 'white'

hyll OE 'hill, elevated piece of land'

hyrst OE 'a hillock a copse'

ing OE connective element linking a first element, a personal name, or a significant word, to a final element.

ingas OE A plural element used in compounded place-names denoting groups of people.

kirkju-by OScand. 'village with a church'

laes OE 'pasture, meadow-land'

land OE 'land'. This element can refer to either a small or extensive area of land.

lang OE 'long'

launde OFr. 'an open space in woodland, a forest glade'

leah OE 'a wood, clearing in a wood'

lin OE 'flax'

lind OE 'lime-tree'

linden OE 'growing with lime-trees'

lundr OScand. 'small wood, a grove, sacred grove'

lytel OE 'little, small'

maed OE 'meadow'

magna L 'great'

market OFr. 'market'

mearth OE 'marten, weasel'

merce OE Mercians, the Anglo-Saxon tribe settled in the West Midlands.

mersc OE 'watery land, a marsh'

middel OE 'middle'

mont OFr. 'mound, hill'

mor OE 'moor, barren waste land'

mos OE 'moss, lichen, bog, swamp'

mus OE 'mouse, field-mouse'

mynster OE 'monastery, the church of a monastery or other religious body'

neothera OE 'lower'

niwe OE 'new'`

north OE 'north, northern'

northman OE 'Norwegian'. This seems to have been used specifically of Norwegian Vikings.

ofer OE 'bank, river-bank, ridge'

ora OE 'border, margin, edge'

oxa OE 'ox'

park OFr 'an enclosed tract of land for beasts of the chase'

parva L 'small'

plot OE 'small piece of land'

pyll OE 'a pool in a river, tidal creek'

read OE 'red'

risc OE 'a rush'

roth OE 'a clearing'

ryge OE 'rye'

sceap OE 'sheep'

sceath OE 'sheath, boundary'. The root of this word means to separate or divide.

scelf OE 'bank, ledge, shelving terrain'

smith OE 'smith, worker in metal'

staca OE 'stake'. Often used to indicate a boundary post.

stan OE 'stone, rock'. This usually alludes to the character of the ground.

stapol OE 'post, pillar'

stede OE 'place, site, locality'

stoc OE 'place, a religious place, a secondary settlement'

stocc OE 'tree-trunk, especially one left standing, stump'

stoccing OE 'piece of ground cleared of stumps'

stow OE 'place, a holy place, a place of assembly'

straet OE 'a Roman road, paved road, urban road'

suth OE 'south, southern'

swin OE 'swine, pig'

svithinn OScand. 'land cleared by burning'

teag OE 'a close, small enclosure'

thistel OE 'thistle'

thorp OScand. 'secondary settlement, dependent out-lying farmstead'

thyrne OE 'thorn, thorn-bush'

ticcen OE 'kid, a young goat'

toft OScand. 'building-plot, enclosure, curtilage'

tot OE 'look-out'

tun OE 'farmstead, village, small estate'

tunge OE tongue'

twi OE 'double, two'

uferra OE 'higher, upper'

under OE 'under, beneath, below'

up(p) OE 'higher up, upon'

vengi OScand. 'field'

vikingr OScand. 'roving pirate, viking'

waeter OE 'water, an expanse of water'

wald OE 'woodland, large tract of woodland, high forest-land'

walh OE 'slave, serf, Welshman, Briton'

weard OE 'watch, ward, protection'

wella OE 'spring or stream'

wenn OE 'tumour'. This might apply to a mound or a burial mound.

west OE 'west, western'

wic OE 'dwelling or collection of dwellings'. Usually with a special purpose such as a farm or a dairy farm. Can also mean a hamlet/village.

withig OE 'willow, withy'

worth OE 'an enclosure'

wrang OE 'crooked or twisted in shape'

wudu OE 'wood, grove, woodland, forest' can also mean 'timber, wood'

yppe OE 'raised place, platform, perhaps a look-out platform'

FURTHER READING

Kenneth Cameron, *English Place-Names*, London 1996.

Barrie Cox, *The Place-Names of Rutland*, Nottingham, 1994.

Barrie Cox, *The Place-Names of Leicestershire, Part 1, The Borough of Leicester*, Nottingham, 1998.

Barrie Cox, *The Place-Names of Leicestershire, Part 2, Framland Hundred*, Nottingham 2002.

Margaret Gelling, *Signposts to the Past: Place-Names and the History of England*, Chichester 1989.

Margaret Gelling and Ann Cole, *The Landscape of Place-Names*, Stamford 2000.

A.D. Mills, *A Dictionary of English Place-Names*, Oxford 1991.

The English Place-Name Society publish countrywide County volumes. A full list of their publications is available from:

> The English Place-Name Society
> Centre for Name Studies
> The University
> Nottingham
> NG7 2RD

Also published by Heart of Albion Press

Rutland Village by Village

Bob Trubshaw

A guide to the history of all the villages in Rutland, with the emphasis on places that can be seen or visited. Based on the author's sixteen years of research into the little-known aspects of the county.

ISBN 1 872883 69 9
2003. Perfect bound. Demi 8vo (215 x 138 mm), 73 + x pages, 53 b&w photos.
£6.95 plus 80p p&p

"Bob Trubshaw's Heart of Albion Press has made a significant contribution to local history publishing in the East Midlands and this latest offering maintains the publisher's reputation for informative books, attractively produced and, importantly, at an affordable price. This A to Z account of the villages of Rutland – a county unsurpassed, in the words of W.G. Hoskins, for its 'unspoiled, quiet charm' – is both readable and very easy to use. Introductory material includes a short outline of Rutland's history and a brief glossary: very useful if you need to check the meaning of architectural terms... "

John Hinks *Leicestershire Historian*

Also published by Heart of Albion Press

Interactive Little-known Leicestershire and Rutland

Text and photographs by Bob Trubshaw

For seventeen years the author has been researching the 'little-known' aspects of Leicestershire and Rutland. Topics include holy wells, standing stones and mark stones, medieval crosses, and a wide variety of Romanesque and medieval figurative carvings - and a healthy quota of 'miscellaneous' sites.

Some of this information appeared in early Heart of Albion publications (mostly long out of print), but this CD-ROM contains extensive further research. The information covers 241 parishes and includes no less than 550 'large format' colour photographs (all previously unpublished).

There are introductory essays, a glossary and plenty of hypertext indexes.

Runs on PCs and Macs.

ISBN 1 872883 53 2
£14.95 incl. VAT.

Special offer!

Mail order customers save 17.5% (because Heart of Albion is not VAT registered) = **£12.70** plus 80p p&p.

Also published by Heart of Albion Press

Sepulchral Effigies in Leicestershire and Rutland

Text by Max Wade Matthews

Photographs by Bob Trubshaw

This CD-ROM makes available for the first time details of the wealth of sepulchral effigies in Leicestershire and Rutland - from thirteenth century priests, thorough alabaster knights in armour and their ladies, to the splendours of seventeenth century Classical aggrandisement. There are even a number of twentieth century effigies too.

350 photos depict 141 effigies in 72 churches, all with detailed descriptions and useful hypertext indexes. Runs on PCs and Macs.

ISBN 1 872883 54 0 **£14.95** incl. VAT.

Special offer!

Mail order customers save 17.5% (because Heart of Albion is not VAT registered) = **£12.70** plus 80p p&p.

Also available from Heart of Albion Press

The History and Antiquities of the County of Leicester

John Nichols

Published between 1795 and 1811, this ambitious antiquarian survey of Leicestershire has long been difficult to obtain. These CD-ROMs contain all the pages of the original editions as high resolution scans.

Published on four CD-ROMs:
    Vol.1 The Town of Leicester
    Vol..2 Framland and Gartree
    Vol.3 East and West Goscote
    Vol.4 Guthlaxton and Sparkenhoe.

Run on PCs and Macs.

Available from Heart of Albion Press for **£10.00** per CD-ROM; **£40** for all four volumes. Please add 80p for p&p.

Also available from Heart of Albion Press

The History and Antiquities
of the County of Rutland

James Wright

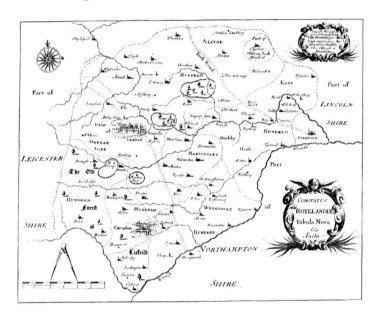

First published in 1684, this pioneering work on local history
contains a wealth of information. However copies of the book are
hard to come by, so the entire text has been scanned in and made
available as high resolution images on this CD-ROM.

Available from Heart of Albion Press for **£10.00** plus 80p p&p.

Full details of current Heart of Albion
publications online at www.hoap.co.uk

To order books or request our current catalogue please contact

Heart of Albion Press
2 Cross Hill Close, Wymeswold
Loughborough, LE12 6UJ

email: albion@indigogroup.co.uk
Web site: www.hoap.co.uk